SUPPLYING WAR

Why did Napoleon succeed in 1805 but fail in 1812? Were the railways vital to Prussia's victory over France in 1870? Was the famous Schlieffen Plan militarily sound? Could the European half of World War II have been ended in 1944? These are only a few of the questions that form the subject matter of this meticulously researched, lively book. Drawing on a very wide range of unpublished and previously unexploited sources, Martin van Creveld examines the 'nuts and bolts' of war: namely, those formidable problems of movement and supply, transportation and administration, so often mentioned – but rarely explored – by the vast majority of books on military history. In doing so, he casts his net far and wide, from Gustavus Adolphus to Rommel, from Marlborough to Patton, subjecting the operations of each to a thorough analysis from a fresh and unusual point of view. The result is a fascinating book that has something new to say about virtually every one of the most important campaigns waged in Europe during the past two centuries. Moreover, by concentrating on logistics rather than on the more traditional tactics and strategy, Dr van Creveld is able to offer a reinterpretation of the whole field of military history.

In this new edition with a new postscript, van Creveld revisits his now-classic text, commenting on the role of logistics in high-tech modern warfare.

Martin van Creveld is a professor in the Department of History at the Hebrew University, Jerusalem. His previous books include *The Rise and Decline of the State* (Cambridge, 1999), *The Sword and the Olive: A Critical History of the Israeli Defense Force* (2002), *Air Power and Maneuver Warfare* (2002), and *Transformation of War* (1991).

SUPPLYING WAR

Logistics from Wallenstein to Patton

Second Edition

MARTIN VAN CREVELD

The Hebrew University, Jerusalem

CAMBRIDGE
UNIVERSITY PRESS

CAMBRIDGE UNIVERSITY PRESS
Cambridge, New York, Melbourne, Madrid, Cape Town, Singapore, São Paulo

Cambridge University Press
40 West 20th Street, New York, NY 10011-4211, USA

www.cambridge.org
Information on this title: www.cambridge.org/9780521837446

First published 1977
Second edition first published 2004
Reprinted 2005

Printed in the United States of America

A catalog record for this publication is available from the British Library.

Library of Congress Cataloging in Publication Data
Van Creveld, Martin L.
Supplying war : logistics from Wallenstein to Patton /
Martin Van Creveld. – 2nd ed.
p.   cm.
Includes bibliographical references and index.
ISBN 0-521-83744-8 – ISBN 0-521-54657-5 (pb.)
1. Logistics – History.   2. Military art and science – History.
3. Military history, Modern.   I. Title.
U168.V36   2004
355.4′11′09–dc22   2003069665

ISBN-13   978-0-521-83744-6 hardback
ISBN-10   0-521-83744-8 hardback

ISBN-13   978-0-521-54657-7 paperback
ISBN-10   0-521-54657-5 paperback

To Louis and Francien Wijler

CONTENTS

MAPS

PREFACE

This study owes its existence to a book, *The Blitzkrieg Era and the German General Staff, 1865–1941*, by Larry H. Addington, which first excited my interest in logistics. Professor Addington has also kindly answered some queries, as did Mr David Chandler and Mr Christopher Duffy, both of the Department of Military History, Sandhurst. I owe a debt of gratitude to Professor Laurence Martin and Mr Brian Bond, of King's College, London, for aid and encouragement. Material for Chapter 7 was made available on behalf of the Liddell Hart Centre for Military Archives by Lady Kathleen Liddell Hart, for whose kind hospitality I am most grateful. Some financial support during my stay in London came from the British Council.

Above all, however, I thank Rachel my wife; it was she who typed out some of the chapters, and suffered for the rest.

London, 28 July 1976   M. v. C.

INTRODUCTION

Logistics are defined by Jomini as 'the practical art of moving armies' under which he also includes 'providing for the successive arrival of convoys of supplies' and 'establishing and organizing. . . lines of supplies'.[1] Putting these together, one arrives at a definition of logistics as 'the practical art of moving armies and keeping them supplied', in which sense the term is used in this study. The aim of the study is to arrive at an understanding of the problems involved in moving and supplying armies as affected through time by changes in technology, organization and other relevant factors; and, above all, to investigate the effect of logistics upon strategy during the last centuries.

Strategy, like politics, is said to be the art of the possible; but surely what is possible is determined not merely by numerical strengths, doctrine, intelligence, arms and tactics, but, in the first place, by the hardest facts of all: those concerning requirements, supplies available and expected, organization and administration, transportation and arteries of communication. Before a commander can even start thinking of manoeuvring or giving battle, of marching this way and that, of penetrating, enveloping, encircling, of annihilating or wearing down, in short of putting into practice the whole rigmarole of strategy, he has – or ought – to make sure of his ability to supply his soldiers with those 3,000 calories a day without which they will very soon cease to be of any use as soldiers; that roads to carry them to the right place at the right time are available, and that movement along these roads will not be impeded by either a shortage or a superabundance of transport.

It may be that this requires, not any great strategic genius but only plain hard work and cold calculation. While absolutely basic, this kind of calculation does not appeal to the imagination, which may be one reason why it is so often ignored by military historians. The result is that, on the pages of military history books, armies frequently seem capable of moving in any direction at almost any speed and to almost any distance once their commanders have made up their minds to do so. In reality, they cannot, and failure to take cognizance of the fact has probably led to many more campaigns being ruined than ever were by enemy action.

Though it has been claimed that civilian historians are especially prone to overlook the role of logistics,[2] the present author has not found this fault confined to any class of writers. Napoleon's tactics and strategy have attracted whole swarms of theoreticians, historians, and soldiers who between them were able to show that both were natural, indeed necessary, outgrowths of previous developments. The one field of Napoleonic warfare that is still believed to have been fundamentally different from anything that went previously is the logistic one, which is itself enough to suggest that the subject has been neglected. Similarly, no one has yet made a detailed study of the arrangements that made it possible to feed an ambulant city with a population of 200,000 while simultaneously propelling it forward at a rate of fifteen miles a day. To take another example: though Rommel's supply difficulties in 1941–2 are probably mentioned as a crucial factor in his fall by every one of the enormously numerous volumes dealing with him, no author has yet bothered to investigate such questions as the number of lorries the Africa Corps had at its disposal or the quantity of supplies those lorries could carry over a given distance in a given period of time.

Even when logistic factors are taken into account, references to them are often crude in the extreme. A glaring instance is Liddell Hart's criticism of the Schlieffen Plan which, while concentrating on logistic issues, does so without considering the consumption and requirements of the German armies, without saying a word about the organization of the supply system, without even a look at a detailed railway map.[3] All we find is a passage about the circumference of a circle being longer than its radii, which reminds one suspiciously of that 'geometrical' system of strategy so beloved of eighteenth-century military writers. And this passage is put forward by some, and accepted by others, as 'proof' that the Schlieffen Plan, the details of which took scores of highly-trained general staff officers half a generation to work out, was logistically impracticable!

Clearly, this will not do. Instead, the present study will ask the fundamental questions: what were the logistic factors limiting an army's operations? What arrangements were made to move it and keep it supplied while moving? How did these arrangements affect the course of the campaign, both as planned and as carried out? In case of failure, could it have been done? Wherever possible, as in Chapters 5, 6 and 7, an attempt is made to answer these questions on the basis of concrete figures and calculations, not on vague speculations. Yet even where, as is often the case, the sources available make it impossible to go into such detail, one can at least analyse the main logistic factors at work and assess their effect on strategy. And one can do this without adhering to stereotypes such as eighteenth-century 'magazine chained' or Napoleonic 'predatory' warfare.

An undertaking to study logistics and its influence on strategy during the last century and a half is very ambitious. To compress the topic into the space of a single book, and yet avoid mere generalities, this narrative concentrates on a number of campaigns between 1805 and 1944 (with an introductory chapter on the seventeenth and eighteenth centuries) selected to present different aspects of the problem. Thus, the Ulm campaign is commonly regarded as the most successful example ever of an army living 'off the country', whereas that of 1812 represents an attempt to utilize horse-drawn transport in order to cope with a problem that was too big to be solved – if it could be solved at all – by anything but the means offered by the modern industrial era. The Franco-Prussian war of 1870, of course, is said to have witnessed a revolution in the use of the railway for military purposes, while 1914 allows a glimpse into the limits of what could be achieved by that means of transportation. The German campaign against Russia in 1941 is interesting as a problem in the transition towards a wholly mechanized army; whereas, in the Allied forces of 1944, that transition had been completed. Finally, Rommel's Libyan campaigns of 1941 and 1942 present some aspects worth studying because unique. From beginning to end, we shall be concerned with the most down-to-earth factors – subsistence, ammunition, transport – rather than with any abstract theorizing; with what success, remains for the reader to judge.

I

The background of two centuries

The tyranny of plunder

The period from 1560 to 1660 has been described as 'the military revolution' and as such was characterized above all by the immense growth in the size of Europe's armies. Marching to suppress the revolt of the Netherlands in 1567, the Duke of Alba made a tremendous impression by taking along just three *tercios* of 3,000 men each, plus 1,600 cavalry; a few decades later, the Spanish 'Army of Flanders' could be counted in tens of thousands.[1] The most important engagements of the French Huguenot wars during the latter half of the sixteenth century were fought with perhaps 10,000–15,000 men on each side, but during the Thirty Years War battles between French, Imperial and Swedish armies numbering 30,000 men and more were not uncommon. At the peak of their military effort in 1631–2, Gustavus Adolphus and Wallenstein each commanded armies totalling far in excess of 100,000 men. Such numbers could not be sustained during the later stages of the Thirty Years War, but growth continued after about 1660. At Rocroi in 1643, the largest power of the time – Imperial Spain – was decisively defeated by just 22,000 French troops, but thirty years later Louis XIV mobilized 120,000 to deal with the Dutch. Even in peacetime under his reign, the French Army seldom fell below 150,000 men, that of the Habsburgs being only slightly smaller, numbering perhaps 140,000. The war establishment of both forces was much larger still, the French one reaching 400,000 during the years of peak military effort from 1691 to 1693. In 1709, it was already possible for 80,000 Frenchmen to meet 110,000 Allied troops on the battlefield of Malplaquet. More and better statistics could be adduced, but they would only serve to prove what is generally recognized: namely, that apart from a period of about twenty-five years between 1635 and 1660, Europe's armies multiplied their size many times over between about 1560 and 1715.

As armies grew, the impedimenta surrounding them increased out of all proportion. Unlike the spruce, well-organized force that Alba took with him to the Netherlands, the armies of early seventeenth-century Europe were huge, blundering bodies. A force numbering, say, 30,000 men, might be followed by a crowd of women, children, servants and sutlers of anywhere between fifty and a hundred and fifty per cent of its own size, and it had to drag this huge 'tail' behind it wherever it went. The troops consisted mostly of uprooted men with no home outside the army, and their baggage – especially that of the officers – assumed monumental proportions. Out of 942 wagons accompanying Maurice of Nassau on his campaign of 1610, no less than 129 were earmarked to carry the staff and their belongings, and this figure does not include a perhaps equally large number of 'extracurricular' vehicles. All in all, an army of this period might easily have one wagon, with two to four horses each, for every fifteen men.[2] Under special circumstances – when it was necessary to try and make a force self-sufficient for an unusually long time, as during Maurice's 1602 campaign in Brabant – the proportion might even be twice as much; on that occasion, no less than 3,000 wagons were collected to accompany 24,000 men.[3]

In view of the ever-growing hordes of troops, women, servants and horses, the methods used to feed them are of some interest. By and large, the military forces of every country consisted of mercenaries; the army as such owed them little more than their *solde*, out of which they were expected to purchase not merely their daily food but also, albeit often helped by an advance from their company captain, their clothing, equipment, arms, and, in at least one case, their powder. Always provided the treasury sent money and that the officers were honest in distributing it, the system could work well enough as long as the troops were stationed more or less permanently in some well-populated place. A regular market could then be organized and put under the supervision of the intendant, who was responsible – to the government, not the commander in chief – for finding out what could be supplied and for policing the premises allocated for commerce and exercising price and quality control.[4] The trade between the troops and the local population was generally conducted on a voluntary basis except when some shortage was expected; in such a case it might become necessary to prevent the richer soldiers from buying up all the available stock for their own use.[5] The system, as is well known, was subject to endless abuses that worked against the interests of almost everyone involved. Nevertheless, there was not in principle anything manifestly impossible about it.

Once the army had to operate away from its permanent station, however, the situation became very different. Establishing markets takes time, and the local peasants could not be counted upon to sustain the force unless, as was usually the case, its movements were slow and marked by lengthy pauses. The prospect of gain could induce some larger merchants – the sutlers properly speaking – to follow the army, but they and their wagons would increase still further the size of its tail[6] while their stocks could not in any case last forever. In friendly territory it was sometimes possible to send commissionaries ahead in order to organize the resources of this or that town and set up a market. In a very few cases, when armies were repeatedly using the same routes for years on end, more or less permanent stations would be organized in which everything required by the soldiers was available for sale.[7] Another method of keeping an army on the move supplied was to quarter it on the inhabitants of the towns and villages on the way. In addition to free shelter, salt and light, these could be expected to provide other necessities *en lieu* of cash payment. In practice, of course, this did not always work out well; as often as not, the soldiers would both take their food and keep their money, not to mention that of their hosts.

On the other hand, no logistic system of the time could sustain an army embarked on operations in enemy territory. Nor, indeed, was the need for such a system felt prior to our period. From time immemorial the problem had been solved simply by having the troops take whatever they required. More or less well-organized plunder was the rule rather than the exception. By the early seventeenth century, however, this time-honoured 'system' would no longer work. The size of armies was now too large for it to be successful. However, the statistical data and administrative machinery which, in a later age, would help to cope with this increase in numbers by turning plunder into systematic exploitation did not yet exist. As a result, the armies of this period were probably the worst supplied in history; marauding bands of armed ruffians, devastating the countryside they crossed.

Even from a strictly military point of view, the consequences of such a situation were disastrous. Unable to feed their troops, commanders were also incapable of keeping them under control and of preventing desertion. To overcome both, but also in order to secure a more regular source of supply than could be afforded even by the most thorough plundering,[8] commanders during the last few decades of the sixteenth century began to see the need to have the army furnish the soldier with at least his most elementary needs, including food, fodder, arms, and sometimes cloth. This, again, was done with the help of sutlers, with whom contracts were signed to supply the army; the resulting expenses were then deducted from the soldiers' pay.[9] The beginnings of this new system can be traced almost simultaneously in the armies of two of the largest powers of the time, France and Spain, led respectively by Sully, Minister of War to Henry IV, and Ambrosio Spinola.[10]

Whatever system of supply was used, the first requirement for a well-ordained army was invariably money. During the second half of the sixteenth century, however, the growth of armies far exceeded that of their governments' financial possibilities. Even the richest power of the time, Imperial Spain, was bankrupted no less than three times by military expense during the period from 1557 to 1598. By the time of the Thirty Years War, no major European State except the Dutch could afford to pay its troops. Consequently, it was necessary to resort to the system of contribution. Though ultimately adopted by all belligerents, it is generally recognized to have originated with Wallenstein, the Imperial commander.[11] Instead of demanding provisions from local inhabitants which were to be paid for by treasury receipts, Wallenstein extracted large sums in cash which then went to the Army cashier, not to the individual soldier or unit. While frankly based on extortion, the system had two distinct advantages: it assured the soldier of regular pay on one hand, and relieved him of the need to rob for his own personal benefit on the other. In intent it was more orderly, and therefore more humane, than its predecessors; though in practice it worked out so terribly that, shocked by its horrors, Europeans everywhere were still making efforts to avoid its repetition a century and a half later.

So much for the supply system of the period. In assessing its effect upon strategy, the most striking fact is that armies, unless they were more or less permanently based on a town, were forced to keep on the move in order to stay alive. Whatever the method employed – whether 'contribution' *à la* Wallenstein or direct plunder – the presence of large bodies of troops and their hordes of undisciplined retainers would quickly exhaust an area. This state of affairs was particularly unfortunate because it coincided with a time when the spread and development of the bastion was rapidly reinforcing the defence as against the offence. If Charles VIII had been able to conquer Italy *'col gesso'*, the strength of a late sixteenth- and early seventeenth-century power no longer consisted mainly in its field army; instead, it lay in the fortified towns, and a country liberally studded with these would even find it possible to wage war without any real field army at all. Under such conditions war consisted primarily of an endless series of sieges; whereas a strategic move into enemy country often struck thin air.

When it came to deciding just which fortress was to be besieged, or for that matter relieved, considerations of supply often played a very important role. The logistics of the age being what they were, a town whose surroundings had been thoroughly devastated might well be immune to either operation. This is well illustrated by the Dutch failure to relieve Eindhoven in 1586, a failure caused less by the difficulty of feeding a force of 10,000 men on its fifty-mile approach march to the place than by the inability to do the same when it was encamped beneath its walls.[12] Since a really protracted siege would cause the surrounding countryside to be completely eaten up regardless of its previous state, it was only possible to conduct an operation of this kind under exceptional circumstances. Thus, Maurice during the siege of Ostend could keep his army supplied from the sea; unfortunately the garrison was able to make use of the same means, the result being that the siege lasted for a record-breaking two years.

In so far as it was possible to eat up one place after another commanders found it easier to operate in the field. Since armies were not supplied from base, and also because in many cases they did not even expect to be paid by the states in whose name they were fighting, lines of communication were of little moment in determining the directions of their movements. The system of contributions made Wallenstein's hordes almost self-sustaining. The same is true of most other forces, including those of Gustavus Adolphus who, from the beginning of 1631 onward, was extracting the bulk of his supplies from the country in a manner not notably different from anybody else's. Except for a few special cases, it was therefore strategically impossible to cut seventeenth-century armies off from anything except, sometimes, their areas of recruitment. Campaigns having this last objective in view were occasionally launched.[13] Subject to the limitations discussed below, armies could – and did – follow the call of their stomachs by moving about freely to whatever region promised supplies, while largely indifferent to their own communications with non-existent bases.[14] Far from calling for speed in operation, this kind of warfare did not even make for a sustained and purposeful advance in any well-defined direction.

As against this almost unrestricted freedom from lines of communication, the strategic mobility of seventeenth-century armies was severely limited by the course of the rivers. This normally had little to do with the difficulty of crossing as such; rather, it stemmed from the fact that the shipping of such supplies as were carried along by water was always very much easier than dragging them overland. While this particular consideration applied equally to all armies it was found, paradoxically, that the better a commander organized his supplies the more dependent on the waterways he became. This was due both to the enormous carrying-capacity of ships as compared to that of horse-drawn wagons, and to the fact that the former did not create additional requirements of their own. Thus, one of the foremost military engineers of the century calculated that 100 *lasten* flour and 300 *lasten* fodder could be contained in just nine ships, whereas on land no less than 600 wagons were needed in order to transport the former alone.[15]

Of all the commanders of the age, none showed himself more adept at exploiting the advantages offered by water-courses than Maurice of Nassau – and, conversely, no one found it more difficult to operate without them. By rapidly shipping his artillery train from east to west and back along the great rivers – Maas, Rhine, Lek and Waal – Maurice succeeded in surprising the Spaniards time and again, appearing now in Flanders, now in Guelderland, always catching the Spanish fortresses before they could be made ready for defence. Once he got away from the rivers, however, he was lost. This is best illustrated from his campaign of 1602, which, incidentally, was one of the very rare contemporary instances of an attempt to win a war by means of purposeful strategic manoeuvre.

Crossing the Maas, Maurice planned to avoid the fortresses on his way, penetrate deep into Brabant, bring the Spanish army to battle and finally swing west into Flanders; the ultimate aim being the liberation of both provinces. For this purpose, a large field army – 5,422 cavalry and 18,942 infantry – was concentrated; he also had thirteen cannon, seventeen half-cannon and five field pieces, but of this artillery train only twelve half-cannon were to accompany the army in the field, the rest being sent by water to meet him. The force was supposed to be self-contained for the first ten days, and was accompanied by 700 wagons carrying fifty *lasten* flour; another fifty were to go by water. In spite of these not inconsiderable preparations, there could be no question of even trying properly to organize the army's supplies for the duration of the campaign; all the above-described measures were only supposed to last the army until it should be possible to harvest the fields on the way and process the grain into bread.

As it was, the campaign was launched too early in the season. Crossing the Maas on 20 June, it was immediately found that the corn of Brabant was not ripe for harvesting. The stores carried along also proved disappointing, the army's English contingent in particular wasting its allocated share and having to be assisted by the others. Maurice thereupon wrote the Estates General that he did not know how he was to continue the campaign, that he would try and bring the Spaniards to battle, but would have to return to the Maas if he was unsuccessful. Having marched for just one week, the army came to a halt on 27 June; for the next three days the process of baking fresh bread was pushed ahead 'with great industry', so that the advance could be resumed on 2 July. When another pause for baking had to be made three days later, Maurice definitely made up his mind that, if unable to force a battle near St Truijen, he would return to the Maas. By 8 July, St Truijen was in fact reached, but then it was discovered that only sixteen out of the fifty *lasten* supposedly following the army by water could be found. Faced with starvation, Maurice decided to retreat. After the remaining flour had been distributed and baked, the march back started on 10 July but had to halt on the next day because it was 'exceedingly hot'. On 12 July, the English contingent had again wasted all their bread and had to be helped out by the army. Back on the Maas, a large consignment of bread and cheese reached Maurice on 19 July, whereupon he determined to march into Flanders. However, the

Estates General had now had enough of his aimless manoeuvring;
they categorically forbade this move, and Maurice settled down
to besiege Grave.[16] It has been claimed that Spain failed to conquer the Northern Netherlands because there were too many
rivers; on their side, the Dutch made no headway in Belgium
because there were not enough of them.

Even those commanders who did not bother overmuch with
the state of their supplies, however, were dependent on the rivers
to some extent because of the enormous weight of the artillery of
the period. For example, in the artillery of Maurice of Nassau,
who among other things was a great artilleryman, the heaviest
pieces – the so-called *Kartouwen* – weighed about 5½ tons and
had to be taken apart for transportation. Even so, they required
no less than thirty horses each, of which perhaps twenty to thirty
per cent were expected to die annually of exhaustion. A modest
artillery train consisting of six half-cannon, each with its 100
rounds of ammunition, required about 250 horses to draw the
guns proper, as well as the wagons loaded with shot, powder,
tools and engineering materials of all kinds.[17] Normally, the
artillery required twice the time to cover a given distance than
the army as a whole, giving rise to complex order-of-march problems on both the advance and the retreat. Not all contemporaries
were content with this state of affairs. Maurice's cousin, Johan of
Nassau, was but one among many who made practical proposals
to lighten the artillery. More important were the efforts of
Gustavus Adolphus, with whom this problem became something
of an obsession. To solve it he abandoned the super-heavy
murbräcker, had the barrels shortened and their thickness reduced, and also introduced a series of ultra-light pieces, the most
famous (if not most effective) of which was the leather gun.
Though these reforms made it possible to reduce the number of
horses and wagons accompanying the artillery by almost fifty
per cent they did not, as we shall see, free his strategy from the
limitations imposed by the relative immobility of the artillery;
nor did they prove lasting, and after his death heavier cannon
were again cast in Sweden.[18]

To sum up, the fundamental logistic facts of life upon which
seventeenth-century commanders based their strategy were as
follows. First, in order to live, it was indispensable to keep moving. Second, when deciding on the direction of one's movements,
it was not necessary to worry overmuch about maintaining con-

tact with base. Third, it was important to follow the rivers and, as far as possible, dominate their courses. All three principles are well illustrated by the career of Gustavus Adolphus, whose operations are generally believed to have been more purposeful than most and are used to demonstrate everything from the importance of having a base to the virtues of the indirect approach. In fact, logistics determined his course of action from the moment he landed at Peenemünde in July 1630. Indeed, were it not for the fact that supply difficulties prevented the Imperialist general, Conti, from concentrating his superior force against him, the landing might not have been possible at all. Even though the King's army numbered only 10,000 men, he found it impossible to feed it in devastated Pomerania[19] and had to expand his base first. To this end he moved this way and that without any apparent strategic aim, taking towns as he went and providing each with a garrison. The process gradually enlarged the area from which he could draw for supplies. However, it also proved self-defeating in that, the more numerous the fortresses besieged or otherwise taken, the more troops had to be found to hold them down. Under these circumstances it is not surprising that he took until the spring of next year before he was able to collect a field army of any size and start operations in earnest.

Having eaten up whatever was left to be had in Pomerania, Gustavus Adolphus felt the need, during the winter of 1630-1, of expanding his base still further.[20] Since not even his artillery could be moved except by water, he had the choice of two routes, either reaching west and southwest in order to get to the Elbe or marching south up the Oder. He tried, but failed, to do the first, and then set out to add Brandenburg to his area of supply. It was during this period that the King, whose promises had induced the citizens of Magdeburg to rebel against the Imperial authority, should have gone to the town's aid. However, he could not do this as long as he did not possess the fortresses of Küstrin and Spandau, respectively guarding the confluences of the Wartha with the Oder and the Spree with the Havel. These all-important waterways could only be secured by negotiations with the Elector Georg Wilhelm. By the time they had been thus secured, however, Magdeburg had fallen.

Almost a year had now passed since the King's landing in Peenemünde, and throughout this period he had lived much as did all other armies, at the country's expense. On 18 July we find him writing to his chancellor, Oxenstierna, from the camp at Werben: 'we have often informed you of our conditions, i.e. that we and the Army are living in great poverty, difficulty and disorder, all our servants having left us, and we are compelled to wage war by ruining and destroying all our neighbours. It is so at this very moment, for we have nothing left to satisfy the men except for what they can rob and plunder...' And in another letter: 'regardless of your own proposal, Mr. Chancellor, to send us 100,000 thalers a month...the Army has not received one penny for the last sixteen weeks...to feed our men we have had only such bread as we could squeeze from the towns, but there is a limit even to that. It has been impossible to restrain the horsemen ...who live simply from wild plunder. Everything has been ruined thereby, so that nothing more can be found for the soldiers in towns or villages.'[21] It was, to be sure, high time for the army to expand its 'base' area once again. In September, the way for this was opened by the signal victory at Breitenfeld.

Having defeated Tilly, two courses were once again open to Gustavus Adolphus. He could continue his way along the Oder; strategically this might have been the logical thing to do, for down to the southeast lay the centre of his enemies' power, Vienna. Alternatively, he might advance to the Rhine. This second route promised better going for the artillery.[22] What was more, it would carry the Swedes into the richest part of Germany instead of the bleak mountains of Bohemia. The upshot was that, once again, logistics were allowed to prevail over strategy. November found the Swedes near Mainz, having in three months taken possession of the greater part of central Germany. Whatever its other merits, the decision had certainly justified itself in terms of the army's material state down to its very appearance. Almost overnight, the crowd of beggars infesting Brandenburg and Saxony had been turned into a rich, well-appointed force.[23] This, of course, was achieved by thoroughly squeezing the towns on the way, priests and Jews everywhere being made to pay an extra contribution.

The winter of 1631–2 was spent in the general area of Würzburg–Frankfurt–Mainz. By this time, Gustavus Adolphus had under his command over one hundred thousand men. In preparation for the next campaign, he hoped to double that number. Even though his troops were now able to draw on the resources of half Germany, however, it was clear that no such enormous

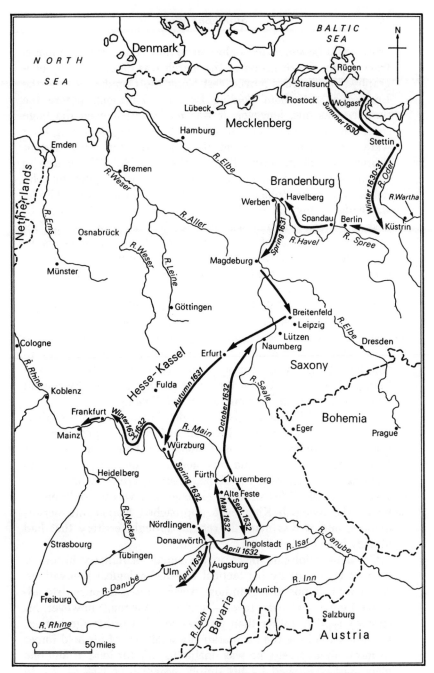

1. The operations of Gustavus Adolphus in Germany, 1630–2

army could be supported unless some new conquests were made. Once again, the direction of the march was determined by geography. The Swedes moved east along the Danube, crossed the Lech, then set about holding Bavaria to ransom. Before the summer was out, however, it was realized that even the huge sums extorted from such towns as Nuremberg and Augsburg were insufficient. To prevent disintegration, the army had to continue its 'flight forward' along the Danube.[24] The march to Vienna was soon interrupted by the news that Wallenstein, by debouching from Bohemia and infesting the Lower Saxon Circle, was endangering the Swedes' communications with the Baltic. In a show of concern that was rare for his day – he was in fact operating hundreds of miles away from his 'base' – Gustavus Adolphus left Donauwörth and, after some marching and counter-marching, arrived at Fürth near Nuremberg where he set up camp. Here the army remained for the next two months, Gustavus Adolphus and Wallenstein each doing his best to starve the other out. As it was, the latter proved more adept at this kind of operation; by early September, the King was forced to march, no matter where. It is a telling comment on the state of the seventeenth-century art of supply that Wallenstein's troops, even though they had just won their first victory over the Swedes at Alte Feste, were too sick and hungry to follow up.

Back on the Danube, Gustavus Adolphus continued his laborious advance eastward into Bavaria. In October, his communications with Sweden were again in danger of being cut, and so he took along 20,000 men and marched for Naumburg in order to seize the crossings over the Halle, covering 270 miles in twenty-seven days. Of all his strategic marches, this was the only one in which the Swedish King even approached Napoleonic performances, and it was made in retreat through territory that had already been occupied and garrisoned.

Gustavus Adolphus' place in the annals of military history is secured above all by his tactical and technological innovations. Apart from impossibly grandiose dreams about a concentric advance by five – or even seven – armies on Vienna,[25] however, his strategy typified rather than transcended that of his age in that, far from exploiting the comparatively small size of his field forces to move about freely, he was unable to stay for long at any one place and compelled by his supply system – or lack of it – to march wherever and whenever his stomach, and that of his horse, led him. With a latter-day monarch, the King might well have said that it was food and forage, not he, which commanded the army. In the next generation, however, efforts to rid strategy of this particular kind of dependence on logistics began to be made.

Rise of the magazine system

'History', wrote Richelieu in his political testament, 'knows many more armies ruined by want and disorder than by the efforts of their enemies; and I have witnessed how all the enterprises which were embarked on in my day were lacking for that reason alone.'[26] Dating to the period when the Thirty Years War was entering its last, and worst, stage, when Central Europe was already so devastated that it could no longer support armies as large as those of the early 1630s, the words well reflect a situation in which, following the departure from the scene of both Wallenstein and Gustavus Adolphus, the problem of maintaining armies of any size was becoming well-nigh insuperable. Tortenson, Baner and Wrangel were never able to concentrate more than 15,000 men at any one point. As the war degenerated into a series of more or less deep cavalry raids against enemy towns, most of which were destined to disintegrate owing to a lack of supplies, it looked as if military art was about to make a return to the middle ages.[27] That it did not end up by so doing was due primarily to the efforts of two Frenchmen, Le Tellier and Louvois. Between them, this pair of father and son established a system of magazines which, during the next century and a half, is said to have exercised a decisive influence upon the wars of the age.

Magazines, of course, had never been entirely unknown. Throughout history, it was frequently necessary to wage war in poor, or ravaged, country. To make sure that designs would not be frustrated by this, military writers of the early seventeenth century advised their readers to set up numerous magazines in conveniently situated towns and fortresses. A well-appointed camp should always have fifteen days' provisions in store, to be touched only in emergency.[28] These and other principles ('never put too many eggs in one basket') are as old as warfare itself. However, the limited size of armies did not make it necessary to put them into practice except for rare occasions.

Like magazines, regular convoys serving to bring up supplies from base were needed only in exceptional cases. Even then, the transport did not form part of the army's establishment; rather, wagons were provided on a makeshift basis, either by commercial contract or, perhaps more frequently, by requisitioning local peasant-carts for which payment was subsequently supposed to be made upon presentation of a suitable receipt. The transportation of supplies, or for that matter of anything else, was apt to be a dangerous business. Centring as it did round fortified towns, seventeenth-century warfare paid scant attention to the gaps between them and did not normally know well-defined 'fronts' to separate enemy from friends. The providing of escorts to protect the convoys was therefore essential under all circumstances and cases are recorded when whole armies found themselves thus employed. Quite apart from the absence of a suitable financial and administrative organization, this factor must have played a major role in the relatively late appearance of a regular system of supply from base. With armies unable to form any kind of continuous defensive lines, or even to thoroughly dominate any large area, a system of this kind was far too vulnerable to cavalry raids.

Appointed intendant to the Army of Italy on 3 September 1640, Le Tellier set out for Turin where he spent the next two campaigns (1641 and 1642) in trying to improve administration. Besides demanding more regular pay for the troops he attempted to combat corruption and, though he did not feel able to dispense with contractors, strove to make them improve on their jobs by imposing stricter contracts, retaining transport through the winter months, and forcing them to keep at hand at least some magazines. These measures were thus directed towards a more effective implementation of existing arrangements rather than towards the creation of entirely new ones. It was only after being created secretary of war in April 1643 that Le Tellier could start rebuilding the supply service in earnest.

The first prerequisite for any regular logistic system is, of course, an exact definition of requirements. So elementary is this point that it may sound self-evident, yet it was necessary for Le Tellier to begin his reforms by laying down in regulations just how much food and other supplies each member of the army was entitled to. The figures varied enormously, ranging from the 100 rations per day allotted to the commander in chief to the single one allowed the private infantryman. Similar provisions were made for the horses carrying or hauling the officers' persons, valets and baggage, the exact number again being dependent on rank. He followed this up by drawing up standard contracts to replace the multitude of *ad hoc* agreements by which the army had hitherto been supplied. Under the terms of such a contract, the war ministry undertook to pay the sutler, exempt his convoys from tolls and other duties, and also to provide them with an escort. Arriving in camp, the sutler was assured a place to ply his trade, protection against excesses by the soldiers and, if necessary, compensation for losses. In return, the sutler undertook to deliver the agreed quantity at the royal depots where its quality would be checked by the *général des vivres*. Responsibility for transporting the provisions from the magazines to the army also rested with the sutler, who was authorized to requisition wagons along the way and pay for them at the normal rate. Having brought the flour into the proximity of the army, the sutler was authorized to engage civilian bakers – by force, if necessary – and make them work 'day and night'. Le Tellier did not therefore depart from existing principles in that he continued to restrict the tasks of the army proper to control and supervision, for which purpose he created a special corps of intendants in August 1643. His greatest single innovation was the establishment of the *équipage des vivres*, a permanent vehicle-park which, guided by specialist army personnel, was designed not so much to carry provisions from the rear as to accompany the army in the field as a rolling magazine with a few days' reserves.

Given these arrangements, the use of magazines gradually became more frequent. As early as 1643 we find Le Tellier amassing provisions at Metz, Nancy and Pont-à-Mousson for the siege of Thionville and to assist Turenne's manoeuvres on the Rhine. In 1644 he established a magazine to feed the cavalry – always the first force to run out of subsistence during a long stay at any one place – while participating in the siege of Dunkirk. Again, in 1648, he told Marshal Gramot to erect magazines at Arras and Dunkirk in preparation for the siege of Ypres.[29] But Le Tellier's greatest achievement came during the very successful campaign of 1658 which, since its course was determined largely by logistic considerations, is worth following in some detail. Turenne left his winter-quarters at Mardic for Dunkirk in mid-May. We are not told how he lived on the way, but it appears that his force had been made self-sufficient during the ten days' march. Arriving at Dunkirk, he laid siege to the town and, after a few days, began to receive supplies by sea from a magazine established at Calais.

The town fell on 25 June. Turenne continued to Brégues which held out for a few days only. He then marched inland, taking one town after another as he went, and having his supply-boats follow him all the time. Early in September he reached Oudenaarde which, again, held out for a few days only. Not having enough food on hand, he was unable to continue to Brussels. Instead, he marched to Ypres which, besieged on 13 September, fell a fortnight later. This conquest enabled him to march up to the Escaut and camp comfortably on its bank. Having wasted some weeks on the river, he made a drive for Brussels in November but found it was already too late in the season. Turenne thereupon saw to it that the places he had taken were fortified and provisioned for the winter, then left for Paris.[30]

Acting on the sound principle that it was possible to supply a campaign – really a series of sieges – only by water, Turenne had thus started the year's work at Dunkirk and subsequently made his way inland. Boats loaded with bread and ammunition followed him wherever he went, and use was made of recently-conquered places in order to set up an expanding network of magazines. The only item for which the army was almost entirely dependent on local supply was, as usual, fodder, and shortages in this commodity did in fact appear whenever it was necessary to undertake a siege of any duration. As it was, it speaks well for Turenne and Le Tellier's organization of the campaign that on no occasion did logistic difficulties force the suspension of a siege, though such difficulties certainly did play a central role in deciding which places should be besieged.

Important as they were, the reforms of Le Tellier belong unmistakably to the age of the 'horde army' in that they bore a temporary, makeshift character. Magazines were established, and stores amassed, in order to support this or that operation; there could be no question of creating any kind of permanent reserve. Indeed, so far was this from being the case that any surplus was always sold immediately after the termination of the campaign, the aim being both to 'relieve the King's subjects' burdens' and to fill their master's pockets. It remained to Le Tellier's son, Louvois, to introduce the first permanent magazines. With that we find ourselves leaving one era and entering another, that of the standing armies.

Louvois, in fact, created not one type of magazine but two. The first of these was supposed to assist in the defence of the realm by designating a chain of frontier towns and fortresses as *places fortes du roi*, which were to be made permanently ready to stand a siege by being always filled with enough provisions to feed the garrison for six months and their horses for two. More revolutionary were the *magasins généreaux*. From these it was intended to meet the requirements of field armies embarking on campaigns beyond France's own borders. Both kinds were placed under governors whose duty it was to see that they should be well stocked at all times, and a good part of Louvois' correspondence contains admonitions to his governors to resist the temptation of using the magazines in order to meet day-to-day needs. Like his predecessors, Louvois as secretary of war supervised rather than negotiated. He did not purchase directly for the State but dealt with contractors, the reason being not merely the absence of a suitable administrative machinery but also a shortage of funds. His procedure as it emerged from innumerable letters and memoranda was generally as follows: first, he would calculate consumption by multiplying the number of troops by the number of days the campaign was expected to last, usually set at 180. Then he wrote down the price he was prepared to pay for each separate item and, by adding the price of transport, storage and distribution to that of purchase, arrived at the total cost. Contracts were then drawn up with the State normally asking for credit. This was perhaps the weakest point of the whole system because, unable to pay the contractors in time, Louvois was helpless in face of their depredations.[31] As it was, the evils involved could not be eradicated and were to last as long as did the *ancienne régime* itself.

Louvois made no innovations in the hauling of provisions from magazine to camp. Regular transportation-corps being still a long way in the future, the normal method remained the employment of locally-requisitioned vehicles supplemented by barges where possible. It was, perhaps, in the system of distribution that Louvois made his most important change. For the first time, it was established as a matter of principle that every soldier was entitled to have his basic daily ration free of charge. Standard fare consisted of two pounds of bread per day, sometimes replaced by the hard biscuit which, a century later, was to serve as the fuel propelling Napoleon's armies to and fro across the European continent. This basic ration was supplemented according to circumstances by meat, beans, or other protein-containing food.[32] These

were not included in the standard fare and were supplied, sometimes free, sometimes at half or quarter their market price. On
one occasion the army as a whole might benefit from 'the King's
generosity,' on others, the infantry only. Concerned with discipline as he was, it did not apparently occur to Louvois to exercise control over consumption, the result being that the men
frequently wasted their food or, alternatively, bartered it for
wine.

Louvois' reforms are said to have allowed increased freedom
of manoeuvre, made possible greater speed in movement and
extended the length of the season during which it was possible
for the French army, and especially its cavalry arm, to stay in the
field.[33] To some extent these claims are indeed valid, as is demonstrated by the first campaign wholly organized by Louvois. This
was also his most successful and, in more ways than one, marks
a revolution in warfare. For Louis XIV's war against the Dutch in
1672, Louvois built what was perhaps the largest field army since
Xerxes, 120,000 men strong, mobilized from all over Western
Europe. The approach to Holland from the south being barred
by row after row of river fortresses, it was decided to launch the
invasion from the east through Guelderland. The army was to be
kept supplied from a chain of magazines established in advance
on the territory of France's ally, the Elector of Cologne, who earmarked four of his towns – Neuss, Kaiserswerth, Bonn and
Dorsten – for the purpose. The next step was to send agents into
Cologne who, while ostensibly working for the Elector, were to
have those magazines filled. The principle followed throughout –
one that may sound strange in these days of 'export or die' – was
to procure from one's own country only that which could not be
got at abroad. In this particular case it was necessary to bring up
the artillery from France, but everything else was acquired outside her own frontiers. The powder and ammunition even came
from Amsterdam by means of the good offices of a Jewish banker
there, Sadoc. At the same time, France's own magazines along her
northern frontier were also filled.

The campaign proper opened on 9 May. Turenne with 23,000
men and 30 guns left his camp at Châtelet (near Charleroi) and
marched down the Sambre. Taking Tongres and Bilsen on his
way, he reached Maestricht and invested it. At this point he was
joined by Condé's Army of the Ardennes which had come marching down the Meuse. On 19 May the united Army left Liège for

the Rhine. Condé crossed the river while Turenne stayed to its left, both marching north along the opposite banks until re-united at Emmerich on 11 June. The Rhine was crossed on 12 June and it was now that the war began in earnest. In just one week the French had reached Amersfoort, sixty miles from Emmerich and perhaps twenty from Amsterdam. At this point the Dutch opened their dykes and, in face of the rising water, the campaign came to an abrupt halt.

From Châtelet to Emmerich the army had marched mainly over friendly territory. Far from drawing away from its magazines, it was in fact getting nearer to them with every day's advance. In spite of these exceptionally favourable circumstances it did not cover more than about 220 miles in thirty-three days, averaging under seven miles a day in what was to be the most mobile of all the campaigns of Louis XIV. It was not until enemy territory was reached that performances somewhat improved, aided by the fact that distances were short and opposition almost nonexistent. It was, to be sure, a highly successful campaign, but one perhaps more marked by the thoroughness of its organization than by any extraordinary mobility.

Coming now to the limitations of the system, it is important above all to keep in mind how incomplete it was. The usual method, indeed the very aim of warfare in this period was to live at the enemy's expense. As Louvois himself put it in a standard formula that he used whenever informing his intendants with the army of a forthcoming campaign, 'His Majesty has. . .a considerable army assembled. . .in order to enter the lands of His Catholic Majesty and live at their cost until the governor of the Catholic Low Countries will give in to his demands. . .[the intendant] will be responsible for making the Spanish lands pay the taxes that will be imposed on them.'[34] Even when it was Louvois' express intention to 'avoid burdening the country', fodder was one commodity that had to be taken away from it on practically every occasion.[35] Usually, however, there was no room for such niceties. French commanders were instructed to 'live off what you can find on the way' even when it was only a question of making a relatively short flank march, and to this purpose were authorized to use means every bit as ferocious as those employed by Wallenstein, including the destruction of houses and the seizure of men, beasts and movables.[36] Even in the early phase of a campaign, when part at least of the army's demand had to be met from a

magazine set up on home territory, 'the means. . .that will enable
[the troops] to live more comfortably' had to be taken from the
enemy and consumed at his expense.[37] That this was a truly
monumental understatement the following few figures will illu-
strate. A typical army of Louvois' day numbering, say, 60,000
men would have 40,000 horses between cavalry, artillery and baggage.[38] At two pounds per head, the men consumed 120,000lb.
bread per day. In addition there were quantities of other food
and beverages weighing, at the very least, another 60,000lb. The
horses' rations could vary very considerably with the season of the
year, but would normally amount to about ten times those of the
men, i.e. twenty pounds per head, making a total of 800,000 a
day for the whole army. Overall consumption would thus amount
to no less than 980,000lb. per day, of which only 120,000 – or just
over eleven per cent – were ever stored in magazines or moved in
convoys. All the rest was procured locally as a matter of course,
either because it was impossible to store and preserve for any
length of time – as in the case of food for men – or because it was
too bulky by far for its transportation to be even remotely possible, which was the case for the horses' fodder.

It is obvious that the need to obtain the ninety per cent of supplies that were not brought up from the rear must have done
more to dictate the movement of armies than the ten per cent that
were, but this has been ignored by the great majority of critics
beginning with Guibert who censured Louvois' 'mania' for magazines, and ending with modern writers who, echoing this criticism,
have described contemporary warfare as being 'shackled' by an
'umbilical cord of supply'.[39] There were instances of the system
of supply from the rear limiting the movements of armies, one of
the most glaring examples being perhaps the inability of Luxembourg to find transport in order to bridge a gap of just sixteen
miles between Mons and Enghien in 1692, yet on the whole it was
the availability or otherwise of *local* supplies, much more than
magazines or convoys, that determined the movements of
Louvois' forces just as it had those of Gustavus Adolphus. This
applied even to the King himself who, together with his 'army' –
that is, a party numbering 3,000 men – could not be assured of
finding provisions on his way to assist at the siege of Luxembourg
in 1684 and had to postpone his departure by two weeks.[40] Reading through the annals of the War of the Spanish Succession, it is
this problem that crops up most frequently. On one occasion

Bourgogne could not reinforce Tallard at Bonn because the latter had barely enough flour to feed his own army while the surrounding countryside had been all eaten up. On another Houssaye explained to Louis XIV that it was impossible to lay siege to Landau because its surroundings had been occupied twice already during previous campaigning seasons and could therefore support neither the besieging nor covering armies.[41] Accused by his royal master of dispersing his troops and thus exposing them to attack, Puységur replied that the Spanish Netherlands were too poor to keep them supplied if they were concentrated in a small area.[42] But on no occasion is the dependence of Louvois' army – the one whose supply system was the envy of all Europe – on the resources of the country brought out more clearly than in June 1684, when Louis XIV was hesitating as to whether his next victim ought to be Mons, Ath or Charleroi. The capture of the former was thought to constitute 'a real blow' to the Dutch, but would present 'invincible difficulties' because there was no subsistence available locally. This being so, Louvois concluded his letter by pointing out to the King that 'it is better to take one of the other places than to do nothing at all'.[43]

If warfare in the '*siècle de Louis XIV*' often appears to us petty and unenterprising, the reason for this was not so much the supposedly exaggerated dependence of armies on their magazines and convoys but, on the contrary, the inability of even the best organized force of the day to do without local supply for practically all its fodder and much if not most of its provisions. From Louvois' correspondence it appears that bodies of troops marching from one place to another beyond the area dominated by France – and sometimes within it as well – were either made self-sufficient for the duration or told to 'live off what you can find on the way'. In *no* instance that I have come across is there any question of a force on the move being supplied solely by convoys regularly shuttling between it and its base, and it has even been claimed that the mathematics involved in this kind of operation were too sophisticated for the military commander of the age to tackle.[44] From beginning to end, the most difficult logistic problem facing Louvois, his contemporaries and his successors was much less to feed an army on the move than to prevent one that was stationary from starving; witness the pride with which the Duke of Marlborough recorded this particular aspect of his operations in front of Lille on the victory column at Blenheim Palace.[45]

Since the system's aim was to sustain armies at times when they were *not* moving – that is, during sieges – the whole image of forces operating at the tip of a more or less extended 'stem' consisting of regular transport – albeit hired or requisitioned – is a false one. Rather, there was likely to be leading to a besieged town not a single, long line of communication but several short ones – normally two to four – each feeding from a magazine containing either bread or ammunition. One for fodder might sometimes be established in order to supplement the resources of the country, never to entirely replace them.[46] Under this logistic system, speed and range counted for very little. Rather, it was a question of disguising and dispersing one's preparations – then as now, dispersion was an essential element of surprise – and of minutely coordinating the movements of troops, siege-train and supplies with their different modes of transportation and controlling them in such a way as to make everything and everybody, from Vauban to the King, appear in front of the selected town at exactly the right moment. It was in demonstrating how to do all this without disturbing normal trade and without arousing the suspicions of the enemy, while overcoming the defective means of communication, administration and transportation of the day, and drawing, when possible, on the enemy's resources rather than one's own, that Louvois' contribution to the art of logistics really lay, not in a futile attempt to endow strategy with any greatly increased freedom of movement.

The age of linear warfare

Of eighteenth-century armies it has been said that they could not march on their stomach but only wriggle on it, and a picture is drawn of forces which, while endowed by their magazines with a certain freedom to choose the direction of their movement, were limited in speed and range by these very magazines and, furthermore, forever concerned with protecting their all-important lines of communication.[47] All authorities agree that the resulting type of warfare was slow and laborious; a few have gone so far as to call it pettifogging and pusillanimous. However, when it comes to analysing just which elements of the logistic system imposed this relative lack of mobility, and how, confusion reigns supreme. Eighteenth-century armies are supposed to have fed from magazines in their rear, yet the very writers who emphasize this point most strongly also claim that the aim of warfare normally was 'to subsist at the enemy's expense', a phrase employed as a matter of course even by the 'prophet of mobility', Guibert himself.[48] It is said that it was impossible for a commander to get more than fifty, sixty, or eighty miles away from their base ('the limit for horse-drawn vehicles') yet all armies seem to have been 'encumbered' by something called 'rolling magazines' which, if the term has any meaning at all, surely implies wagons loaded with provisions which, by making the army self-contained for a time, should have allowed it to move in any direction, to any distance, for as long as stores lasted. The picture, then, is contradictory. Clarification is called for.

Apart from everything else, the character of eighteenth-century warfare was a direct consequence of its political ends. Wars were regarded as personal feuds between sovereign princes; one had a claim or a 'grievance' against one's neighbour, to obtain satisfaction of which one sent an army into his territory and lived at his expense until he gave in.[49] If there was any hope of permanently keeping the province in question after the war, care was taken to preserve one's future capital; if not, exploitation could be ruthless. In any case, magazines would only be established to provide the army with an initial push. It would then cross the border, march into enemy territory, and select a convenient place – easy to defend, situated near a good road or river – to set up an entrenched camp. The procedure to be followed from this point onward is well described by that incomparable writer on eighteenth-century warfare, Maurice de Saxe: 'provisions will not be lacking for present consumption, but some management is required in the method of procuring supplies for future exigencies. . .In order to accomplish this it will be necessary to fall upon a method of drawing supplies of provisions and money from remote parts of the country. . .the best way is to transmit to these places. . .circular letters threatening the inhabitants with military execution on pain of their refusal to answer the demand made from them'.[50] Parties numbering twenty to thirty men, commanded by an officer, were sent to gather the contribution, failing which, the magistrates' houses could be either plundered or burnt down as appropriate.

The money having been collected, a combination of force and persuasion was used in order to purchase or commandeer supplies at prices suitably fixed by the army's intendant. While the presence of an enemy in the neighbourhood might make it expedient to speed up this process in order to prevent him from drawing near,[51] it was normally carried out without hurry. One simply stayed in one's place until the resources of the surrounding country were exhausted. The first commodity to go short was invariably fodder, and when this happened one gathered whatever supplies were still to be had, struck camp and moved to another area.

While the strategy of 'eating all there is to eat'[52] did not require any very great administrative resources – its very purpose, indeed, was to make it possible to wage war without such resources[53] – problems were bound to occur whenever a siege lasted longer than expected. To capture a town before the resources of the surrounding country gave out was a cardinal problem of warfare, and to solve it the curious (if logical) arrangement under which the terms obtained by a garrison stood in inverse proportion to the length of its resistance was instituted. Even in the eighteenth century however, such expedients did not always work, and when they failed it became necessary to set afoot those tremendous supply-operations that have become the laughing stock of all subsequent critics. To mention but one example, and that the most famous of all, in 1757, Fredrick II was compelled to raise the siege of Olmütz because an indispensable convoy with 3,000 vehicles was intercepted by the Austrians. Later in the year, therefore, we see him taking great precautions to avoid a similar disaster. First, 15,000 men were used to escort a convoy from Tropau to Olmütz, then 30,000 to bring another to Königgrätz, and finally 8,000 to secure the line of communications to Glatz. It was, in Clausewitz's words 'as if the whole Prussian war-engine had ventured into the enemy's territory in order to wage a defensive war for its own existence'.[54] That this was so on this occasion is undeniable. The implication that Frederick's army was too cumbersome to move, however, is false, because this state of affairs was caused, and could only have been caused, by the demands of siege warfare.

That Frederick II could move very fast when he wanted to is proved by his performances in the field. In September 1757, he took thirteen days to cover 150 miles from Dresden to Erfurt; two months later, he covered 225 miles from Leipzig to Parchwitz in fourteen days. Again, in September 1758, he took just one week to march 140 miles from Küstrin to Dresden and a year later he covered 100 miles from Sagan to Frankfurt on Oder in a week in spite of having to fight the battle of Minden on the way. As Clausewitz himself points out, Frederick did away with his baggage and supply convoys in order to carry out these marches; nor were they needed, for an army moving from one place to another could always find something to eat.

This brings us to the so-called 'five days' system', the methodical shuttling to-and-fro of wagon columns that is supposed to have exercised a paralysing influence on eighteenth-century strategy. As described by Tempelhoff, the limiting factor consisted above all in the number of flour-wagons operating between the field-bakeries and the magazines in the rear; since 1/9 of these were emptied every day, the maximum distance that an army could get away from its base was 9:2=4.5 marches, or approximately sixty miles. Even Tempelhoff himself, however, admits this to be an underestimate, conceding that it was possible to extend the range by another forty miles by using the regimental bread-wagons to carry flour as well. In fact, the entire system was a myth, the clever invention of an armchair strategist. Even Frederick II, the only commander who is said to have tried it out in practice, helped himself by other means whenever the opportunity presented itself.[55] Nor, it should be remembered, was the system complete even in principle, for only a very small percentage of an army's needs was ever supplied from base. In particular, the problem of transporting the huge quantities of fodder required was so difficult that its solution was never even attempted.[56]

How relatively simple it was to organize the supply of an army on the march even when the distances involved were considerable is proved by Marlborough's famous movement from the Rhine to the Danube in 1704. This was not a very large affair by contemporary standards; the Duke had with him just 30,000 men and was joined by another 10,000 on the way. The practical details of this march are well known. Setting out at dawn, the troops would cover twelve to fourteen miles each day and reach camp at noon. The cavalry went ahead of the infantry and the artillery which, in the charge of the Duke's brother, got stuck in bad weather and consequently fell well behind schedule.[57] Nevertheless, between 20 May and 26 June, some 250 miles as the crow flies were covered, though the actual marching distance must have been closer to 350. After this performance, men and horses were still in sufficiently good condition to win Prince Eugene's unstinted admiration.

How was it all done? It has been said that 'the essence of Marlborough's system of transport and supply...consisted of a contract with one Sir Solomon Medina who supplied it with bread and
bread wagons'.[58] This is an oversimplification. The Duke had
originally hoped to steal a march on his enemies by establishing
magazines and preceding them into the field,[59] but money for
this was not provided by the Dutch Estates General and nothing
came of the project. Nor could such magazines have provided
him with more than an initial push, since problems of transportation and preservation made it all but impossible for their contents
to be carried very far. Instead, Marlborough got his bread from
the brothers Medina who in turn purchased it in the surrounding
country, employing no doubt local agents. Everything else the
troops were expected to buy from their pay, for which purpose
each company and regiment had individual contracts with sutlers
great and small. We are thus presented with a charming picture.
The troops arrive in camp around noon each day and are greeted
by the sutlers with soup kettles at the ready. The local peasants
have also been alerted and are happy to sell their products to
soldiers who, for once, were well able to pay their way. One eats
one's fill, settles accounts, then goes on siesta.

Reality, of course, was different. Small though it was, Marlborough's army was too big to simply buy its way forward.
Rather, it was necessary to make sure in advance that enough
supplies would be ready and available for purchase. To this purpose, the Duke would send ahead impeccably polite letters, e.g.
to the Elector of Mainz: 'it would please your highness...to see
to it that we may find provisions on our way, pending prompt re-
payment. It would be very advantageous for the troops and also
for the country in preventing disorders if [this]...may be arranged by sending officers ahead to regulate everything...'.
Similarly, he informed the Estates of Franconia that he had sent
ahead *munitionaires* of the army to collect supplies and asks for
cooperation.[60] When such cooperation was not forthcoming the
consequences could be dire. Marlborough would express his
'surprise' at the resulting disorder, then inform the recalcitrant
town that, since they were now presumably more ready to accept
his 'protection' than before, he was sending a detachment for
the purpose with orders to scour the surroundings and bring in
everything edible for man and beast, a task in which the magistrates were kindly asked to assist.[61] Marlborough's advance thus

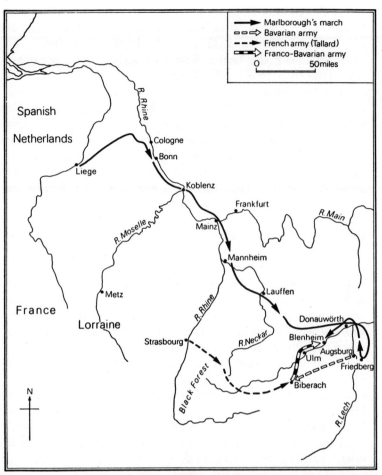

2. The Campaign of 1704

swept the country bare, with the result that he, like Wallenstein
before and Napoleon after him, was unable to traverse the same
country twice. This caused him to split the army into detachments
when following the beaten Franco-Bavarian forces back westward after their defeat at Blenheim.[62]

As regards supplies other than food, Marlborough followed a
similar procedure. Rather than taking the trouble to accumulate
stocks, he simply purchased whatever came his way, hiring transport to bring the goods into camp when possible, marching the
army or part of it to fetch them at the place of manufacture when
necessary.[63] Far from constructing magazines to support a war of
movement, Marlborough did so only when the army *stopped* moving around. Thus, we find him writing from the camp at Aicha in
Bavaria, where he had arrived after completing the march from
the Netherlands, that 'we came to this camp on Friday, and have
since been drawing what corn and provisions we could for erect-
ing a magazine in this place, intending to leave a garrison here. . .'.
And again, from another camp at Friedberg: 'since we are going
to stay here for some time, take care about the magazines'.[64] Nor
were these isolated incidents; rather, they formed part of a
strategy which, late in July and early in August 1704, was moti-
vated almost exclusively by logistic considerations.

The aim of Marlborough's operations during this period was, in
his own words, to prevent his opponent – the Elector of Bavaria
– from getting the least subsistence out of his country.[65] Coming
from the northwest, he beat the Elector at Donauwörth and
forced him to retreat across the Danube and southward along the
right bank of the Lech at Augsburg, where he made use of the
town's fortifications to establish himself safely. The Duke also
crossed the Danube, and marched along the opposite bank of the
Lech until reaching the above-mentioned camp at Friedberg.
Having thus interposed himself between the Elector and his
country, he set out to ravage Bavaria, sending out flying columns
which, after taking away everything movable, set fire to the rest.
Marlborough himself described this operation, sometimes as
coercive, sometimes as punitive – intended either to force the
Elector to defect on his French alliance or to punish him for not
doing so.[66] The real object, however, was military. Faced with a
numerically superior enemy – the Elector had now been reinforced by Tallard's French – Marlborough preferred to starve him
out rather than risk a battle. He therefore intended to move

against Ingolstadt and Ulm, from where the Franco–Bavarian troops were drawing their supplies. He did not, however, feel strong enough to carry out this move to the north and west of his opponents while still leaving a force to guard the Lech. It was necessary to find some other means to make the crossing of the river not worth while for his opponents. Having achieved this by plundering Bavaria, Marlborough retraced his steps along the Lech and returned to the Danube. Rather than sitting still at Augsburg and watching the noose of starvation draw tightly around their necks, however, Tallard and the Elector raced the Duke to the Danube and, by occupying a strong position at Hochstadt, blocked his way. Instead of depriving the enemy of supplies, Marlborough was now faced with starvation once he got through the provisions taken from Bavaria. No help could be expected from the Empire, and though 'it was thought a very hazardous enterprise to attack such a numerous army, as they were so advantageously posted',[67] there was little to be done but risk all. Thus, the battle of Blenheim was fought and won.

By exercising careful foresight and sometimes accompanying his requests for supplies with barely disguised threats, Marlborough was able to feed his army without much difficulty. Problems only arose when he halted at any one place, in which case fodder was invariably hard to come by.[68] On the move, however, there was no need for a complex logistic apparatus to encumber his troops and prevent them from making marches hardly inferior to those of Napoleon one hundred years later. The comparative rarity of such marches should be attributed, not to their impracticability but to their ineffectiveness in an age when a state's main strength consisted in its fortresses. The battle of Blenheim did not, after all, end the war against France, while the march of Marlborough and Prince Eugene against Toulon proved to be a blow in the air.

The relative ease with which it was possible to feed an army on the move also explains why it was unnecessary to establish a regular supply corps. Proposals in this direction were sometimes made,[69] but met with no reaction; in spite of all abuses, the rulers of the age were unanimous in considering it cheaper to employ contractors whose main advantage was that they could be dismissed after the war was over. The contractors, however, never supplied more than a fraction of the army's requirements; in particular, fodder invariably had to be gathered on the spot in complex, well organized operations.[70] The troops of the time being 'the scum of the earth, enlisted for drink', these foraging expeditions led to much desertion, and it was to prevent this that the Austrian army set up the first supply-corps in 1783. Far from being expected to help bring up supplies from base, the task of this corps was to collect them from the country.[71]

That the density of Europe's population, and the development of its agriculture, made it perfectly possible to feed an army as long as it kept on the move the following figures will show.[72] An army of, say, 60,000 men, required 90,000 bread rations per day.[73] At a baking ratio of 3:4, twelve ounces flour were needed to produce a pound of bread. Assuming a daily consumption per head of two pounds the total quantity of flour needed in a ten day period was $\dfrac{90,000 \times 3 \times 2 \times 10}{4} = 1,350,000$lb, or 600 tons. At a density of forty-five people per square mile[74] and assuming a region was self-sufficient, the quantity of flour available in April – when campaigns normally started – must have amounted to six months' supply, or probably around $180 \times 2 \times 45 = 16,200$lb, around seven tons. Given a strip of country 100 miles long and ten wide – which meant that foraging parties had to go no further than five miles on either side of the road – the total quantity available must have been around 7,000 tons, of which less than ten per cent would be needed to feed the army during its ten days' march-through. Problems, therefore, could only arise when it became necessary to stay for long in any one place; in other words, during a siege.

More difficult, but essentially not dissimilar, was the provision of fodder. Even upon the most pessimistic assumption, an acre of green fodder could feed fifty horses.[75] The 40,000 animals accompanying an army would therefore require 800 acres per day. Taking ten days to cross a strip of country ten miles wide by 100 long, an army's horses would therefore devour the fodder grown on 8,000 acres, i.e. 1/80 of the total area.[76] That a much larger area must have been devoted to growing fodder, however, is easy to prove, for 8,000 acres would only have fed 4,000 horses for one year,[77] a number which, compared with that of the inhabitants, is much too low.[78] Lacking exact figures, it is not possible to say just how long an army was able to stay in any given area. All we know is that, whenever a siege was conducted or a camp maintained for long, fodder was invariably the first commodity to run out.

Since only a fraction of the victuals needed for man and beast were ever available in ready-made form, however, an army staying for long in any one place was turned, inevitably, into a food-producing machine that milled grain, gathered wood, baked flour, and reaped fodder. Since the tasks involved had to be carried out rhythmically every few days, it is easy to see how much they interfered with the regular functioning of an army.[79] Indeed, it was perfectly possible for an army's military function to be virtually suspended in favour of logistic ones for entire periods. This was especially the case when fodder had to be reaped, and extraordinary precautions to avoid being surprised during such a time were necessary.

Before drawing our conclusions from these facts, a word must be said about the consumption of ammunition in the wars of the two centuries preceding the French Revolution. That figures on this problem are so very hard to obtain is itself an indication that it was not very important. Compared to the provision of subsistence, that of ammunition was to remain insignificant until well after the Franco–Prussian War of 1870.[80] Indeed, so small were the quantities required that armies normally took along a single supply for the entire campaign, resupply from base being effected only on comparatively rare occasions – most frequently, of course, during sieges.[81] In the first half of the seventeenth century, an army going on campaign took along a basic load of 100 balls for each artillery barrel, which is not surprising in view of the fact that, even during a siege, no gun was expected to fire more than five times a day.[82] Later in the century, Vauban calculated four rounds per gun per day, so that consumption of ammunition remained negligible beside that of food and fodder.[83] The figures for field operations were lower still; during two battles in 1636–8, the Bavarian artillery only fired seven shots per gun in eight hours, though these figures owe their survival to the fact that they were regarded as a record low.[84] Frederick II, who relied heavily on his artillery, usually took along 180 balls on campaign, and instances when a shortage of ammunition forced him to alter his plans only occurred during sieges. Apart from this, there is no evidence that the problem of ammunition supply had any influence on the conduct of operations, nor is the issue mentioned by even those exponents of eighteenth-century warfare who most strongly insist on magazines as a *conditio sine qua non* for a successful campaign.

'An umbilical cord of supply'?

In the annals of military history, late seventeenth- and eighteenth-
century strategy is often said to occupy a special place. Opinions
as to the exact nature of its characteristics vary. To some, it was
a 'civilized' age when the ideas of the Enlightenment and the
decline of religion as a motivating force allowed some humanity
to be introduced into the business of war, to others it was the
period of 'limited warfare' and 'strategy of attrition', both of
which meant that campaigns supposedly aimed not so much at
the complete overthrow of the enemy as at the attainment of
definite politico-economic goals, by making it more expensive for
him to continue the war than to make concessions.[85] To another
school, the limits of warfare in this period resulted not so much
out of choice as from necessity, and among the factors contribut-
ing to these limits, logistics are usually allowed pride of place.

Given the enormous effect that 'the shackles of supply' and
the 'tyranny of logistics' are supposed to have exercised upon
strategy in this period, it is a curious fact that research into the
actual methods by which armies were supplied and kept moving
has not to date advanced far beyond what was known 150–200
years ago. As we saw, it was Tempelhoff who invented the 'five
days' system' that subsequently came to be ridiculed in virtually
every book on the subject, and it was Clausewitz who, more than
anybody else, emphasized the difference between Napoleon's
logistics and those of his predecessors. Between them, these two
writers have been allowed to dictate what is to this day the
'accepted' view on the movement and supply of eighteenth-
century armies.

A detailed examination of the logistic methods employed, however, shows the whole of this picture to be absolutely without
foundation. *Pace* Tempelhoff, not even Frederick II ever employed the 'five days' system' except perhaps during the first
three years of the Seven Years War, and even then the limitations
inherent in the technology of the age made it impossible for him
to transport much more than ten per cent of the army's requirements. Clausewitz notwithstanding, eighteenth-century armies
lived off the country as a matter of course and at least one of them
– the super-scrupulous Habsburg army – even organized a special
supply corps with precisely this objective in mind. As a consequence, eighteenth-century armies were capable of marching-

performances much better than is usually acknowledged;[86] ironically, some of these marches are spelt out by Clausewitz himself. That armies did not usually cover more than ten miles a day for any length of time is true, but what force, its men limited to their own feet and unmetalled roads, has ever been able to do more?

Contrary to the impression given by many modern books on military history, whose authors seem to delight in calculating the size of the 'tail' needed to feed the armies of a Louis XIV or Frederick II, such descriptions are rare in contemporary sources. The scale of baggage allowed – especially to the officers – was admittedly generous, but, this had nothing to do with the need to supply the army from base.[87] The only case I have been able to discover when a commander actually explained his inability to carry out an operation by the size of the supply columns occurred in 1705, when Houssaye told Louis XIV that, to besiege Landau, a train fifty-four miles long would be needed.[88] Nor was this due to any weak-heartedness or reluctance to live off the country; rather, Landau had already been besieged twice in previous years (by Tallard in 1703, by Marlborough in 1704) so that the surrounding area was thoroughly exhausted.

As for supply convoys accompanying armies in the field, there is little in contemporary sources to suggest that their presence inevitably acted as a brake on the movements of armies, nor that they could not be got rid of when the occasion demanded. In any case, it was primarily when sieges were undertaken that such convoys were needed at all; and it was only then that there could be any question of a regular resupply from base. For the rest, eighteenth-century armies lived as their predecessors had always done, and as their successors were destined to do until – and including – the first weeks of World War I; that is, by taking the bulk of their needs away from the country.

That eighteenth-century armies did not, in comparison with their successors, show much expertise in the art of living off the country may be true, but this was not due to any excessive humanity. What was missing was an administrative apparatus specifically responsible for feeding them in the field. As a result, foraging had to be engaged on by the troops, and since this was rightly thought to cause heavy desertion it could only be carried out in complex, highly-organized operations. Wary of the latter, commanders were usually content to leave the exploitation of the country to the contractors, whose depredations were such that armies could starve even in the richest territories.[89]

If long, rapid marches were comparatively rare, and if military critics insisted on the need to base any campaign on long prepared magazines, this stemmed not from any inability – much less, unwillingness – to live off the country but from the fact that late seventeenth- and eighteenth-century warfare concentrated above all on sieges. This in turn was due, partly to a military system (the standing army) that made soldiers too expensive to be lightly risked in battles. partly to a conception which regarded war primarily as a politico-economic (as distinct from moral or ideological) instrument aimed at the attainment of definite, con- crete objectives, and partly to the fact that, unless the enemy de- cided to stand and fight, a strategic march deep into his country was likely to hit thin air. Add to this the extraordinary power of a town fortified by Coehorn or Vauban, and the fact that for- tresses are by nature unable to run away, and the reasons why sieges were so frequent, and rapid marches so relatively few, will become apparent.

Had an army of, say, 100,000 men, wanted to bring up all its supplies for the duration of the campaign – usually calculated as 180 days – from base, the resulting burden on the transportation system would have been so great as to make all warfare utterly impossible. Calculations as to the quantities that would be needed in such a case were sometimes made,[90] but, there is no indication that they ever constituted more than a theoretical exer- cise. To even imagine that the huge quantities of fodder that were required by the 60,000 horses which accompanied such a force could ever be brought up from base borders on the ridicu- lous.

That logistics did exercise an influence on the two centuries of warfare under discussion is of course true, but this influence had little if anything to do with supposed 'umbilical cords of supply' restricting the movements of armies. Rather, the problem facing commanders from the Thirty Years' War onward consisted pre- cisely in that armies, with their hordes of retainers, had grown too large to remain for long at any one place; like the Flying Dutchman, they were forever doomed to wander from one loca- tion to another. Under such conditions, some of the most famous marches of the age – notably that of Prince Eugene from the Tyrol through Venice into Lombardy in 1701 – really consisted of 'flights forward' and were carried out for no better reason than that, since money for magazines was not forthcoming, it was impossible to sit still.[91] Furthermore, these conditions turned every one of the many sieges that had to be undertaken into a race against time. It was in order to solve this problem, not to secure any increased mobility, that Le Tellier and Louvois first set up their system of supply magazines. In this they were successful, yet we have seen that the magazines never contained, nor could contain, more than a fraction of the army's needs, and that French troops on the march continued to live off the country as a matter of course.

Finally, in one sense, the whole concept of supply from base was contrary to the spirit of the age, which always insisted that war be waged as cheaply as possible – an age, indeed, when wars could be launched for the sole purpose of making the army live at one's neighbour's expense rather than one's own. To imagine that rulers as parsimonious as Frederick II ever took away from their own country a single thaler that could possibly have been stolen from another is to misunderstand, not merely the eighteenth century but the very nature of that horrible and barbaric business, war. This fact, no mere 'enlightenment' has to date done much to correct.

'An army marches on its stomach!'

The end of siege warfare

It may be that no fanfares greeted Napoleon's decision of August 1805 to go to war against Austria. However, in the annals of military history, his campaign marks the transition from the eighteenth to the nineteenth century. Brilliant and even novel as the future Emperor's early exploits may have been, they belonged, in more ways than one,[1] to the previous age. In 1805 his strategy changed, and what subsequent critics were to term *Vernichtungsstrategie* was born and came of age at one and the same moment. It is not easy to say just what constituted Napoleon's new style of warfare; too many factors were involved for a simple definition to be possible. Much has been written about the 'new' and 'democratic' character of the French army of the time, whose impregnation with a revolutionary ideology endowed it with an altogether new type of driving force, and about Napoleon's single-minded concentration on the defeat of the enemy's army instead of the attainment of geographical objectives, and his determination to wage war *à outrance*, to break his opponents' heads rather than chopping off their limbs. That there is much truth in every one of these explanations none would deny, yet they seem to ignore the central point; namely that, in order to adopt a 'new' system of strategy, it is necessary first of all to find new means to carry it into practice. It is above all on these means that the present chapter will concentrate.

Owing to the enormous size of the forces under his command, the logistic problems presented by Napoleon's new strategy were of an order of magnitude that was altogether new. We have already tried to show that long strategic marches were, even in the eighteenth century, by no means as impossible to carry out as is usually assumed. If they were invariably small scale affairs, this cannot be attributed primarily to logistic considerations. Rather, the difficulty was that, whereas the relationship between offence and defence was such that one field army could engage another of equal size with reasonable hope of success, a numerical superiority of no less than 7:1 was thought necessary in order to besiege a strong, well-defended fortress. Therefore such an operation could only be engaged on by a large force, e.g. the 120,000 men that Marlborough concentrated round Lille in 1708. When such forces were not available, however, the *only* thing that could be done was to march and fight – multiplying numbers by velocity, no doubt, but even so without much hope of achieving a decisive victory as long as the enemy's fortified places remained intact. Between 1704 and 1712, Marlborough and Prince Eugene followed up one victory in the field with another, yet France's power remained unbroken, and she ended by making a favourable peace.

If Marlborough with 40,000 men marched to the Danube, therefore, and Eugene with 30,000 to Toulon, this reflected not merely their 'casting off the tyranny of logistics' but, to an even greater degree, the fact that to use such small forces against even a single first-class fortress was a hopeless undertaking. Given such small armies, indeed, it was impossible even to merely invest fortresses while simultaneously pushing forward. This, more than any supposed dependence on supply from magazines, explains why such strategic marches as were embarked on normally skirted the enemy's territory instead of penetrating deeply inside it.[2] Had Marlborough been forced to seal off, much more capture, even one strong fortress on his way from Flanders to the Danube the whole of this famous strategic manoeuvre would surely have come to nothing.

It was, in fact, his inversion of the relationship between sieges and battles – between the relative importance of the enemy's fortresses and his field army as objectives of strategy – that constituted Napoleon's most revolutionary contribution to the art of war. Early in the eighteenth century, Vauban counted 200 (unsuccessful) sieges but only 60 battles during the previous two centuries.[3] Napoleon, however, conducted only two sieges during his entire career, and his experiences in feeding the army round Mantua demonstrated that solving the logistic problems of siege-warfare was far from easy even for a Bonaparte. As he wrote to his stepson in 1809, 'the method of feeding on the march becomes impracticable when many troops are concentrated'. Hence it was necessary, in addition to requisitions from the immediate neighbourhood, to have supply convoys coming in from places farther away. That 'this [combination] is the best method'[4] Maurice de Saxe would have readily agreed.

While Napoleon's determination to concentrate all his troops at the decisive point led to his insisting that garrisons should not be provided by the field army but made up of the local population or the national guard, this does not mean that he regarded fortresses as entirely without value. His armies, however, were large enough to invest a fortress – more than one, if necessary – *and* to continue on their way forward. As he once wrote while commenting on Turenne's maxim that an army should not exceed 50,000 men, a modern force of 250,000 could afford to detach a fifth of its number and yet remain strong enough to overrun a country in short order.[5] Far from abolishing fortresses altogether, therefore, it was merely necessary to have them relocated in new places. Instead of fortifying the frontier (which he regarded as folly, since it meant exposing one's depôts and centres of arms-manufacture) they had to be located deep in the rear, preferably round the capital.[6] Carrying out a siege under such conditions would inevitably cause logistic difficulties, as the Germans – the railways notwithstanding – found out, to their cost, in 1870.

Napoleon, in short, realized that it was the eighteenth-century predilection in favour of siege warfare that led to endless logistic difficulties. As he himself was able, thanks to the size of the forces under his command, to do without sieges he rendered the logistic apparatus of the eighteenth century largely superfluous. Hence the explanation for a fact which, at first sight, appears incomprehensible. Although the technological means at his disposal were by no means superior to those utilized by his predecessors – indeed, he was rather conservative in this field, rejecting new inventions and doing away with some old ones – Napoleon was able to propel enormous forces right across Europe, establish an Empire stretching from Hamburg to Sicily, and irreparably shatter an entire (old) world. How it was done, the following example – concentrating as it does round the campaign of 1805, his most successful ever – will try to show.

Boulogne to Austerlitz

The amount of attention paid to bureaucratic and logistic detail

was not, as is well known, one of the strongest points of the Republican armies that came to be established after 1789, and
though the French army had come a long way since Napoleon
found it, to use his own words, 'naked and ill fed. . .among the
rocks', its administrative organization in 1805 was still by no
means impressive. All questions of administration were the
domain of the Ministry of War Organization, whose head at this
time was Dejean. Responsible among other things for feeding,
dressing and equipping the army with transport, the minister's
authority ended at France's frontier. In the field, responsibility
for administrative matters – supply and transportation included –
fell on the army's intendant-general whose powers, however, were
strictly limited to the zone of operations. While administration
and supply at both ends of the pipeline were thus well regulated,
it was typical of contemporary warfare that there existed no permanent machinery to control the zone of communications or exploit its resources. Here the Emperor would make *ad hoc* arrangements, usually by thrusting responsibility upon commanders
whose achievements in the field he deemed unsatisfactory and to
whom employment on such a task was therefore something of a
reprimand if not an actual punishment.[7]

In 1805, the Army's intendant was Petiet. Under him came
four war-commissionaries and the heads of the various branches
of supply (*régisseur des vivres-pain, des vivres-viande, de
fourage*, and the *directeur général des équipages de transport*).
Napoleon, however, constantly bypassed this central organization
by sending orders regarding transportation and supply directly to
his corps commanders. The latter had on their staff an *ordonnateur* whose job it was to look after the corps' supplies in accordance with the broad directives laid down by the intendant at
Imperial Headquarters. Each divisional staff also included a
commissionary who received his orders partly from the corps
ordonnateur and partly from his direct superior, the divisional
commander, two authorities that could, and sometimes did, come
into conflict.

The material means at the disposal of these officials were totally
inadequate. This had nothing to do with any lingering Republican
tradition of hungry, marauding hordes living off the country –
since the Peace of Amiens, Napoleon had had ample time to correct this state of affairs had he wanted to – nor to any misconcep-
tions as to the importance of a proper train. Rather, it stemmed

from the fact that the army had spent the last year and a half preparing for a landing in England. Given the superior British strength at sea, this operation had no hope of relying on a regular line of communication leading back to the Continent. Once it got across the Channel by one means or another, the French Army would have had to live off the country – which, after all, was rich enough to support it – while its hopes for returning to France would have rested entirely on a rapid victory in the field and, following this, a dictated peace. The army at Boulogne was therefore almost totally without any means of supply and transportation. Even if we assume that Napoleon's project for invading England was never more than a feint, however, he could not have created the machinery for waging war on the Continent without alarming his enemies. But, as was always the case, a major consideration governing Napoleon's logistic arrangements for the Austerlitz campaign was the need, at all cost, to preserve operational surprise.

Thus, the campaign of 1805 faced the Ministry of War and the army's Intendanture with the gigantic task of hurriedly scraping together the entire transport and supply apparatus for 170,000 French troops in the space of a few weeks; a problem made all the more formidable by the fact that, during all this time, the bulk of the units for whom this apparatus was destined were not stationary but marching from their camps round Boulogne to the deployment area on the Rhine. An additional complication was presented by the 80,000 new recruits whose organization into units had to be completed even as the process of deployment was in full swing. Indeed, the fact that it telescoped the normally separate stages of mobilization and deployment into a single, combined operation was not the least unorthodox aspect of Napoleon's plan of campaign.

Having decided, on 23 August, to go to war against Austria, Napoleon's first move was to order his forces to their areas of deployment. Of his eight corps, two – commanded by Marmont and Bernadotte – came respectively from Holland and Hanover and, forming the Army's left wing, were to concentrate first at Göttingen and then at Würzburg. The remaining corps were originally supposed to deploy along a line stretching for some fifty miles from Hagenau northward through Strasbourg – where more than three corps were to be concentrated – to Schelestadt,[8] for which purpose five of them had to cross the entire width of France from west to east, while the sixth consisted of recruits arriving in bits and pieces from every corner of the Empire. The march towards the area of deployment thus presented problems of coordination and supply on a gigantic scale. It speaks volumes for the efficiency of Napoleon's chief of staff, Berthier, that he could start sending out his orders on 25 August and report on them to the Emperor just 24 hours later. And these orders were detailed; they laid down not merely the sequence of formations but also the exact quantities of provisions that each regiment was to draw from each individual place on the way.

Under Berthier's master plan, the divisions forming the cavalry corps were the first to leave the Channel coast, which some of them started doing as early as 25 August. Then followed the infantry corps of Davout, Soult, Ney and Lannes marching along three parallel routes arranged from north to south so that only the last two had to share a road between them. Recording Napoleon's instructions, Berthier wrote that 'it is the Emperor's intention. . .to feed the troops on the route as they were in camp', i.e. by means of the machinery provided by the Ministry of War in coordination with prefects, sub prefects and mayors on the way. Provisions were to be distributed every two or three days, and as he issued his orders to the corps Berthier also wrote to the various local authorities informing them of the coming movement and asking for their cooperation. Marching from Holland over friendly territory, Marmont was told to 'live off what the country can supply'; whereas Bernadotte was to take along seven or eight days' supplies of biscuits in order to 'avoid burdening' the neutral country of Hesse-Kassel.[9] Within this general framework much initiative was left to the marshals, each of whom was to send his *ordonnateur* and commissionaries ahead in order to make the detailed arrangements.

Excellent as Napoleon's orders may have been in principle, the organization of the march left something to be desired, both because the time allowed was too short and because local authorities everywhere were reluctant to cooperate with the army which, thanks to its system of recruitment, was already becoming unpopular. On the other hand, there were cases when too much cooperation led to drunkenness and disorder. Davout's corps in the north found it difficult to secure appropriate quarters and had to spend more than one night in the open. The right-hand corps fared better in this respect but suffered when their supply-system broke down while on the final leg of their march. Apart from a phenomenon of 'temporary desertion' – men exploiting the fact that they were passing in or near to their homelands in order to slip away for a few days and then rejoin their units on the Rhine – discipline was excellent and moved at least one prefect to write Berthier that he had 'nothing but praise' for the troops passing through his department. Others were less content, and on 11 December – almost five months since the beginning of the campaign and a week after it had been crowned by the victory of Austerlitz – the Minister of Finance was still complaining to Dejean about the army's failure to settle accounts with the districts through which it had been passing in August and September.[10]

On the whole, however, the men did well enough on the march; there are records left of extreme fatigue,[11] but not of hunger. The same could not be said of the horses which suffered heavily from bad roads, rain and a shortage of fodder. Nearly every commander who depended on the animals to carry his troops or haul their equipment had cause for complaint, either because the riders were untrained and caused injuries to their mounts, or because the horses were too young and broke down, or simply because there were not enough of them. By the time the concentration on the Rhine was complete the horses of the cavalry corps were starving because there were no funds to buy fodder. Soult had only 700 animals to draw his train instead of 1,200 he needed, whereas Marmont's cavalry suffered from the effects of having been cooped-up for five weeks aboard ships in preparation for the abortive invasion of England. It was the old story that we shall see repeated right up to 1914; in every case feeding the horses and keeping them in good health proved a good deal more difficult than doing the same for the men.

Exactly what Napoleon's operational plans were at this stage we do not know. In his letters to Talleyrand, his foreign minister, and to the Elector of Bavaria, his ally, he did not go beyond generalities – marching to the aid of Bavaria at the earliest possible moment, sending 200,000 men to Vienna, and beating the Austrians before they could be reinforced by the Russians. How and in what manner he hoped to achieve these objectives the Emperor did not say, possibly because he was determined to maintain secrecy but more probably because he did not yet know himself. Nor are there any real clues as to his intentions in the instructions he gave for the conduct of reconnaissance in Germany. Thus, Bertrand was ordered to carry through a thorough survey of Ulm and its surroundings, then to proceed east along the left (north) bank of the Danube while paying special attention to the exits through which the Russians might debouch from Bohemia. Murat, presumably in preparation for the operations of Marmont and Bernadotte, was instructed to follow the Main to Würzburg, then to reach the Danube, which he was to descend to the Inn. Travelling south along that river, he was to reach Kufstein before turning west and, after crossing Bavaria, to return to France by way of Ulm–Rastadt which he was to reconnoitre with special care. A better understanding of the Emperor's intentions can perhaps be obtained by looking at his orders of deployment. These initially included a very heavy concentration at Strasbourg, where Petiet was supposed to prepare tents for no less than 80,000 men, or enough to house almost half of the entire *Grande Armée*. Add to this the fact that Napoleon asked his Bavarian ally to prepare large stores of provisions at Ulm, and it becomes fairly clear that, whatever his ultimate purpose, the Emperor's first intention was to forestall the Austrians in Bavaria by marching there over the most direct route, the one through the Black Forest. The fact that he did not have this area reconnoitred does not gainsay this conclusion. The Black Forest was after all, the route traditionally followed by French armies during the wars against the Habsburg Empire, and its features must have been well known.

While Napoleon was pondering his operational plans and supervising the vast process of deployment and concentration, Dejean, Petiet and Murat – the latter acting in his capacity of commander of the army in the Emperor's absence – were making great efforts to complete all material preparations within the short time allowed. Dejean's ordeal began on 23 August, when a curt order from the Emperor instructed him to prepare 500,000 biscuit rations at Strasbourg and another 200,000 at Mainz, all to be ready within twenty-five days. A similar if more polite note went to the Elector of Bavaria who was to prepare no less than a million biscuit rations equally distributed between Würzburg and Ulm.[12] A comparison of these figures with the number of troops involved shows that Napoleon's preparations were by no means as sketchy as is usually supposed. The 700,000 rations concentrated in the area of deployment would have lasted the 116,000

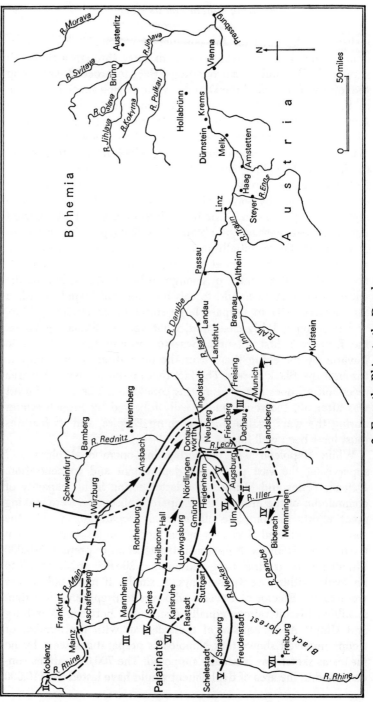

3. From the Rhine to the Danube

men that formed the main body of the *Grande Armée* (excluding the corps of Marmont, Bernadotte and Augerau; the latter was to take no part in the early stages of the operation) for six days, which together with four days' bread would easily have taken them to Bavaria. Here, another four days' supply was supposed to be waiting at Ulm. Preparations for feeding the northernmost corps were just as extensive, arrangements being made to have supplies to last 55,000 men (including the 20,000 Bavarians whom Marmont and Bernadotte were to pick up at Würzburg) for no less than nine days in addition to the normal four days' bread. All in all, the army's subsistence was thus to be secured for a full two weeks, which was more than enough to enable them to reach Bavaria without making any requisitions at all.

As it was, meeting all of the Emperor's demands did not prove possible. In the area of deployment proper, only 380,000 biscuit rations, or just over half of the number originally demanded, could be made ready by 26 September. Another 300,000 were ready in the rear but could not reach the army in time for the beginning of operations.[13] The Bavarians failed to prepare anything at all either in Ulm – which, in view of the surprisingly rapid Austrian advance to that place, was just as well – or at Würzburg, and on 15 September they had to be asked 'for heaven's sake' to prepare at least 300,000 rations in Würzburg, the response being that it could not be done because prices were high and biscuits unknown to the bakers. Nevertheless, it appears that when Marmont and Bernadotte arrived at Würzburg they did find some biscuits awaiting them, though how many we do not know.[14]

While the preparations made to secure provisions were thus falling behind the Emperor's expectations his instructions for providing the army with adequate transport also proved incapable of fulfilment. According to the original plans, the army's vehicle park was to consist of the following: (*a*) 150 wagons brought up from Boulogne; (*b*) just under 1,000 wagons to be furnished by the Compagnie Breidt, with whom a contract had been placed in May; and (*c*) 3,500 wagons to be requisitioned in the French departments along the Rhine. The total number of wagons supposed to accompany 116,000 men – Marmont and Bernadotte had been warned that they could 'expect nothing' from the army and told to find their own transport – may seem meagre by modern standards. It was less so at the time, and in fact the ratio of vehicles to men happened to be exactly the same as in the Austrian army under General Mack.[15] Again, the 2,000 wagons left to the *service des vivres* after the artillery had taken its share of 2,500 would have sufficed to carry provisions for 116,000 men for eleven days if one assumes that consumption amounted to three pounds per day and that the average load of a four-horse wagon was around one ton – which is probably too low. As it was, only a small part of this transport materialized. The wagons from Boulogne were sent to the wrong place owing to a bureaucratic error, whereas the Compagnie Breidt only had about a fifth of its vehicles ready in time. Finally, the drivers of carts impressed in large numbers along the Rhine used every opportunity to desert, taking their horses along when possible.

If the *Grande Armée* suffered from a shortage of transport, this was due to the lack of time in which to set up proper trains rather than to any preconceived determination on the Emperor's side to do without them. Exactly when he first became aware that it would not be possible to meet all his demands we do not know but a clue may perhaps be found in a letter to Dejean in which, on 28 August. he ordered the 500,000 rations previously assigned to Strasbourg to be distributed between that town, Landau and Spires. What is even more significant in our context is that there is no mention at all made of the 200,000 rations ordered for Mainz, nor do they reappear in any subsequent document. Immediately after this reduction in the total amount of provisions an order was sent to Savary to reconnoitre, for the first time, the crossings of the Neckar in an area well to the north of that which Murat had previously been instructed to traverse. Finally, on 30 August, there came a whole series of orders altering the army's area of deployment and shifting it sixty miles to the north so that it now extended from Strasbourg through Hagenau to Spires with a strong concentration of forces on the left wing. In accordance with these new dispositions the bulk of Napoleon's forces would now move through the rich territories of Baden and Württemberg instead of by way of the Black Forest whose defiles had witnessed the sufferings of many a previous French army. Discussing the reasons behind this change, Alombert and Colin claim that it could not possibly have been due to logistic causes because the poverty of the Black Forest must have been well-known to Napoleon long before he gave his original orders. If, however, one agrees with our above conclusion that the Emperor never intended his troops to live by requisition while on their way to Bavaria, and that he first perceived the inadequacy of the means at his disposal during the very last days of August, this objection falls by the wayside.

The area of deployment having been changed, preparations to receive the approaching columns of the *Grande Armée* continued. Towards the middle of September Murat returned from Germany and began inspecting the work done, sending back enthusiastic reports to the Emperor. On 17 September he was at Landau, on the next day he visited Strasbourg, and three days later he spoke to Petiet who, Murat said, was fairly bursting with enthusiasm and sure he could lay hands on everything down to the last detail. This feeling was shared by few of the marshals as they started arriving on the Rhine. On 22 September Soult informed Murat that 'whatever the intendant general may say' his troops at Landau were in danger of running out of bread, while there was nothing left to be purchased in the country around. On the next day, Davout presented Petiet with a list in which he expressed his dissatisfaction with every single detail of his corps' material preparations.[16] The shortage of transport in particular was such that barely enough vehicles to carry the army's ammunition were available, the result being that Marmont had to send forty per cent of his ammunition by way of the Rhine. Arriving at Mainz, it was found that there was not enough water in the Main for the barges to continue their voyage to Würzburg, and so the corps marched on minus much of its ammunition and heavy artillery.

Napoleon himself had meanwhile arrived in Strasbourg where he reprimanded the officers responsible for the army's logistics ('carry out the orders I gave you', he snapped at Dejean) and busied himself with the details of marching and feeding the army once it had got across the Rhine. His orders for the corps were laid down on 20 September. Davout, Soult, Ney and Lannes were to cross the Rhine on 25 and 26 September, each marching by a separate route according to the usual Napoleonic system. Marmont and Bernadotte were also given their marching orders, the latter being ordered to cross neutral Prussian territory at Ansbach in order to avoid congestion. However, he got into trouble when the Prussians, who had originally granted passage for thirty-five days, suddenly revoked their permission and so caused the French marshal to be separated from his heavy baggage which remained

stuck at Hanover. The same orders also provided for each corps
to take along bread for four days and biscuits for another four,
the latter to be regarded as a reserve and touched only in an
emergency. As it was, few of the marshals succeeded in taking
out these strongly reduced quantities and even Davout, who was
accused by Soult of creating for himself a reserve for seven or
eight days at his (Soult's) expense complained that he did not
have enough. Mutual recriminations among the marshals and
complaints to the Emperor were to mark the campaign from beginning to end, yet one cannot fail to recognize the vigour of the
young commanders and their readiness to assume responsibility
in leading tens of thousands of men hundreds of miles away from
France with nothing but an incomplete supply apparatus to support them.

Once across the Rhine, Napoleon kept his corps well apart and
provided for them all, except the southernmost one, to live off the
country to their left,[17] an arrangement which must have caused
some hardship as units had to find quarters at distances greater
than those that might otherwise have been necessary, but which
was probably meant to overcome the shortage of maps and allow
the troops to forage without friction. The details of the operations
were fixed by each corps separately and the instruction issued by
Ney can be taken as a model of its kind. The normal method of
subsisting the men was to quarter them on the inhabitants together with their horses. Rations for men and NCOs were fixed at
1½lb. bread, half a pound of meat, one ounce rice (or two ounces
dried fruit) per day, while wood for cooking was also to be supplied by the unwilling hosts. No exact scale of rations for the
officers was laid down, it being determined merely that they were
to be 'decently fed in accordance with their rank' but without
making 'excessive demands' upon the inhabitants. Whenever
the corps' troops were too closely packed together for this method
to be practicable the *ordonnateur* was responsible for com-
mandeering supplies from the neighbouring areas. He and the
divisional commissionaries were to inform local authorities of the
number of men and horses to be fed and the demands made on
each of them, as well as fixing the place or places to which the
provisions were to be brought. No payment for anything was to
be made, but receipts specifying the exact quantities appro-
priated were to be handed out in all cases so as to make it possible
for the French treasury to settle accounts with the State authori-

ties at some unspecified future date.[18] In issuing these orders, Ney did not forget to tell his men that they were to treat the inhabitants as if they were French. In theory at any rate the *Grande Armée* was therefore very far removed from a mere host of marauding plunderers. Rather, its supply system resembled that of Marlborough in that provisions were accumulated in advance along its route, the only difference being that the Duke paid in cash and not in paper receipts. As to the actual method of procurement, we have already seen that Marlborough could be quite as ruthless, though perhaps more polite, as his successor one hundred years later.

When the Rhine had been crossed, the bridges over it were closed in accordance with an order of 29 September which directed that all traffic to and from the army should pass through Spires. The town was put under the command of an officer (General Rheinwald) who was thus made responsible for the upper part of the pipeline leading to the army. Every five to six leagues (fifteen to eighteen miles) a relay station was set up, while the line of communications was policed by auxiliary troops of the Baden army as well as brigades of gendarmes. Through these points streamed reservists and convoys, and also sick, wounded and prisoners on their way back to France. The original line of communication was the road from Spires to Nördlingen, but on 5 October the commander of Spires was made responsible for the entire area to the right of the Rhine and ordered to see to it that all transports to and from the army went through Heilbronn. At this stage the line of communication still ended at Nördlingen which served as an advance base from where distributions were made to the corps.

While Napoleon's orders were admirable, their execution was less so, particularly because of the ubiquitous shortage of transport. Whatever the corps could raise from the country they naturally retained for their own use, the result being that the line of communication services desperately lacked wagons and horses. The cavalry and the corps wagon-masters stole and hid all the animals they could lay hands on, and by 11 October it was no longer possible even to maintain a regular courier service with the homeland. Napoleon thereupon intervened in his usual decisive manner and ordered the corps to give up their surplus transport.

Starting as they did from a front that was well over 100 miles wide, the corps ought to have advanced with little friction and in fact did so apart for some minor incidents – for example, an error of Berthier's almost sent Davout across Soult's line of march on 30 September. Two days later the routes of Lannes and Ney crossed each other as the former pushed from Stuttgart towards Ludwigsburg. Lannes' corps was so unfortunate as to share a road with d'Hautpoul's cavalry division preceding it and the Guard following in its wake. Repeatedly, Murat complained that Ney was poaching on his preserves. Bernadotte, whose original itinerary included the city of Frankfurt, was turned away from there at the last moment, the result being that he had to reach Würzburg by circuitous roads and so exhausted his troops that they had to be granted three days' rest upon arrival. Difficulties were also experienced in feeding the artillery park making up the army's rear, with the result that convoys had to be sent to it from Spires. On the whole, however, the army did well enough during the first ten days. Soult, Lannes and Ney all made large requisitions, and Davout was able not merely 'to live very well' off the country but to build himself a reserve of six to nine days in addition to the 200,000 biscuit rations following his corps of 25,000. Complaints only came from the two corps on the extreme left wing, whose difficulties probably stemmed from the fact that Napoleon's orders for biscuits to be baked at Würzburg had been only partly obeyed. At the same time, the Prussians refused to sell anything to Bernadotte's men crossing their country.[19] It was to Bernadotte, too, that Berthier wrote on 2 October: 'as to subsistence, it is impossible to feed you by magazines...the entire French army, the Austrian army even, lives off the country'.

As they gradually turned requisitioning into a fine art, the corps *ordonnateurs* were able to draw enormous quantities of supplies from the towns and villages on their way. Thus, for example, Soult forced Heilbronn and its surroundings – total population perhaps 15,000–16,000 – to surrender no less than 85,000 bread rations, 24,000lb. salt, 3,600 bushels of hay, 6,000 sacks of oats, 5,000 pints of wine, 800 bushels of straw and 100 four-horse wagons. Hall and its district probably had only about 8,000 inhabitants but were nevertheless made to yield 60,000 bread rations, 35,000lb. meat (seventy oxen), 4,000 pints of wine, 100,000 bundles of hay and straw, 50 four-horse and 100 other carts, as well as 200 horses with harness. Even much smaller places were able to bring forward truly astounding quantities. For example, Marmont and his 12,000 men stayed for five days in the village of Pfhul (forty houses, 600 inhabitants) and 'did not lack for anything'.[20] Bugeaud's famous question to his sister, 'judge for yourself if 10,000 men arriving in a village can easily find enough to eat', should be answered in the affirmative!

As the army approached the Danube, however, the situation suddenly worsened, probably reaching its nadir round 9–12 October but slowly improving thereafter. Many factors were responsible for this state of affairs, especially the fact that the enemy was now close at hand which made it impossible to prepare stores in advance. Instead of being centrally organized, the task of requisitioning devolved on the divisional commissionaires and sometimes even on individual regiments. This in turn led to every unit fending for itself, with the cavalry in particular racing ahead and occupying the villages that had been earmarked for the infantry and making it almost impossible for the latter to find any supplies at all. Occasionally instances of gross mismanagement occurred; for example Marmont complained about the 'scandalous behaviour' of one Baron Lienitz, master of the Circle of Wasstertrudingen, who had caused 20,000 biscuit rations passing through his territory to be held up. Most important, however, was the fact that the *Grande Armée* was now operating in a relatively restricted area. Its original front, which on 30 September had stretched over more than 100 miles from Freudenstadt to Würzburg, had contracted to just forty-five on 6 October.

While much has been made of the difficulties afflicting the army during this period, one should not lose sight of its achievements. For example, considerable Austrian stores were captured at Memmingen, Friedberg, Augsburg, Donauwörth and Saldmünchen where they had apparently been piled up to await the arrival of the Russian army. Though no corps had been ordered to take along stocks of fodder we find a total of ninety-eight wagons loaded with that commodity passing through Heidenheim on 7 October, of which fifty-four belonged to Ney alone; at one ton each, this should have been enough to feed his 2,600 horses for at least two days. At Hall, Davout had requisitioned fodder for no less than thirty days, and on 10 October Dumas informed the Emperor that he had just seen 3rd Corps' 'beautiful train' with stores for six days passing through Neuburg. Nor is it entirely true, as is so often claimed, that Napoleon's armies could only live as long as they kept moving fast. Although considerable difficulties were caused by the corps settling down to besiege Ulm or to form a strategic barrier round Munich and Dachau it did enable the system of supply to be put on a more organized footing,[21] the result being that provisions became more plentiful after about 20 October.

Throughout the march from the Rhine to the Danube most questions of supply had necessarily been the responsibility of the individual marshals, Napoleon himself being able to contribute little but reprimands ('general Marmont had an order to obtain four days' bread and have four days' biscuit baked; he cannot count on anything but his own resources') or exhortations to improvise, replace one commodity by another, and secure the troops' provisions 'by hook or by crook'. Though 'forced by circumstances' to march without magazines he fully realized the danger of such a procedure, and accordingly began his efforts to put the logistic apparatus on a more sound basis even before the fall of Ulm. On 4 October the Emperor decreed the establishment of a second line of communication back to Spires in order 'to have a secure service of subsistence by this route.[22] On 12 October all the corps were ordered to disgorge their surplus of requisitioned transport and put it at the disposal of the artillery park. Twelve days later the Emperor issued instructions for the establishment of a huge base near Augsburg, the aim being to concentrate no less than 3,000,000 rations – enough to feed the army for 18 days – within the next fortnight. In addition, the marshals were at this time drawing provisions for themselves from such towns as Munich, Ingolstadt, Landshut and Landsberg. These efforts resulted in even Ney, whose corps was not normally noted for the excellence of its organization, obtaining a twelve days' supply. To transport these quantities the Emperor appears to have relied on the arrival of the vehicles promised by the Compagnie Breidt, as well as a flotilla of barges then being organized at Augsburg.

Meanwhile, the service of communications back to France was also being vastly expanded. An order of 23 October divided the line from Strasbourg to Augsburg into seventeen sections, each of which was to be covered by sixty four-horse wagons shuttling to and fro. Assuming that each wagon was capable of making a return journey each day – not an exaggerated requirement in view of the distances involved[23] – the overall capacity of the service must have amounted to between sixty and 120 tons per day consisting mainly of clothing and ammunition. These arrangements may appear sketchy by today's standards. At the time, however, they represented a triumph of organization in making possible the maintenance over unprecedented distances of a regular system of supply and transportation such as Marlborough marching over the same country a hundred years previously neither wanted nor needed. While the quantities involved were admittedly small, they were clearly regarded as ample at the time – the proof of this being that, far from expecting a shortage, the same order made explicit provisions for dismissing some of the transport should there be too much of it.

As to the *Grande Armée*'s ammunition supply, we saw that, in the eighteenth century, the latter exercised little or no influence on the strategic movements of armies owing to the very small quantities involved. Napoleon, however, allocated 2,500 out of 4,500 wagons with which he intended to equip his forces to the artillery park – which also carried two thirds of the supply of infantry ammunition – and only 2,000 to the *service des vivres*. A typical division of 8,000 men took along 147–300 rounds per gun but only 97,000 infantry rounds in addition to the sixty to eighty carried on each man's back.[24] While the quantities involved were thus by no means negligible in either relative or absolute terms, the fact that they could *not* as a rule be procured in the theatre of operations prevented them from becoming a brake on strategy. Like his predecessors, Napoleon was taking along at the outset most, if not all, the ammunition required for the duration of the campaign. Far from being indifferent to this aspect of his logistic requirements, the Emperor was in advance of his time when he established, immediately after Mack's surrender at Ulm, a great artillery depot at Heilbronn through which went a daily flow of 75,000–100,000 rounds of ammunition. This is perhaps the first example of a continuous resupply of ammunition ever recorded, and together with the previously mentioned relay service suggests that, far from reverting to a more primitive logistic method, Napoleon's system was one link in the chain of developments that was ultimately to make modern armies truly shackled to an umbilical cord of supply.

Resuming its advance down the Danube, the *Grande Armée* engaged on what was effectively a completely new operation. The main enemy at this stage was no longer the Austrians but the Russians, and the aim of the advance, not to beat an opponent

who stayed put in his place, but to try and come to grips with one
who, though he might occasionally put up a rearguard action,
was forever retreating and slipping away, threatening to draw
Napoleon behind him into the endless spaces of Bohemia, Poland,
and elsewhere.[25] The 'mechanical' problem presented by this
retreat was all the more formidable because the nature of the
terrain did not allow wide outflanking movements in the best
Napoleonic style but instead tended to funnel the army into the
narrow space between the Danube and the Alps where the number of roads was constantly decreasing. There were five roads
leading from the base on the Isar to the Inn, three from there to
the Enns, but only one from there on to Vienna. Attempts to find
additional roads through the mountains farther south failed and
the corps which tried to get through (Davout's) got stuck in the
mud. Consequently, numerous corps had to share a single road,
columns reached monstrous length, and the army as a whole
tended to lose its cohesion as well as the ability to concentrate the
corps. Throughout this march the Emperor thus found himself
racked between the desire to catch the Russians and the fear that
the advance guard might become engaged against superior forces
while out of touch with the main army, a dilemma which he did
not succeed in solving and which was finally to compel him to go
way beyond Vienna to Austerlitz.

Having been instructed to procure themselves an eight days'
supply of bread and biscuit in Bavaria, the corps forming the
Grande Armée crossed the Isar in three columns with forty miles
between them on 26 October. Murat, Davout and Soult marched
in the centre, forming a column fifty miles long; Lannes was on
the left, Bernadotte on the right. It was intended, after the initial
supplies gave out, to feed the corps once more by means of
orderly requisitions carried out by the *ordonnateurs* and paid for,
even though this was enemy country, by receipts. To this purpose
an attempt was made to allocate each corps a separate foraging
area, Davout, for example, being ordered to leave the country to
his right untouched so that it could be utilized by Soult following him. Marmont, who from 27 October was marching behind
Bernadotte, was told to 'go as far as necessary' to his right in
order to secure supplies from an area already traversed by the
latter. Some of the marshals, notably Davout, Lannes and Soult,
also had convoys from Munich following in their wake, though
the speed of the advance was such that there was no hope for

them to catch up with the troops over roads which, in addition to being crowded, were also covered by snow and ice.

Logistically as well as strategically, the army's march from Munich to Vienna may be divided into three stages. The first of these brought the French to the river Inn. As the country in between was wooded and extremely poor[26] the troops must have crossed it by consuming the stores requisitioned from the inhabitants of Bavaria who, as the Bulletin of 28 October euphemistically put it, had demonstrated 'great zeal and industry' in catering to their needs. From the Inn to the Enns the country opened up and it was possible to make considerable requisitions. As the Inn had served as Mack's original area of concentration, some Austrian magazines were also captured at Braunau, Altheim and Linz. During the first few days of November many divisions could consequently be fed by means of regular requisitions at the hands of the corps *ordonnateurs*. Far from any grave shortages arising, some units at least seem to have enjoyed a surplus which the troops, strict orders to the contrary notwithstanding, either sold or simply threw away.[27] Finally, there came the week-long march from the Enns to Vienna. It was at this stage that serious difficulties arose and complaints similar to those already heard around Ulm were voiced once again. However, this had nothing to do with Napoleon's system of supply or lack of it; since four or five corps were now crowded on a single road, problems were bound to appear whatever the method adopted. Yet the only alternative was to cross the Danube and advance along both its banks, which was a dangerous thing to do in view of the fact that all the bridges had been burnt by the retreating Russians and any force operating north of the river thus became exposed to being caught in isolation. In the event, so bad did conditions along the road to Vienna become that Napoleon decided to take the risk, which led directly to the best part of a division being annihilated in the 'affair' of Dürnstein.

This reverse that the *Grande Armée* suffered was due partly to the fact that the Emperor allowed a sixty-mile gap to develop between himself and the advanced spearheads, which in turn was caused by his determination to personally supervise the organization of his supplies at Linz. Thus, far from being 'indifferent' to his communications, he allowed his care for them to interfere with his conduct of operations. The establishment of an intermediate magazine at Haag was ordered as early as 29 October. On the same day the Emperor also provided for a depôt to be set up at Braunau, the aim being to bake there 50–60,000 rations per day in preparation for an expected stand by the Russians. When the desired combat did not take place it was decided to turn Braunau into an advanced centre of operations. Flour for three million rations was to be concentrated there and baked into bread at the rate of 100,000 rations per day. Transportation to the army was to be carried out both by water and overland, for which Soult, Ney and Bernadotte were ordered to disgorge surplus transport. Meanwhile, the equivalent of three allied divisions were detailed to guard the line of communications.

Important as these measures were, they were destined primarily to form a hedge in case of failure and could hardly contribute much to sustain the operation which, as Murat informed the Emperor, was beginning to assume the character of a 'flight forward.'[28] Happily for Napoleon, Vienna was now near at hand. Here, there were such vast quantities of arms and ammunition ('enough to equip three or four armies', as the Bulletin put it) that all the *Grande Armée*'s difficulties in this respect were solved at one stroke. More important still, 10,000 quintals of flour and 13,000 bushels of fodder were found in the Imperial magazines alone. The town was ordered to find food to last 80,000 men for three weeks, which involved the delivery of 75,000lb. bread, 25,000lb. meat, 200,000lb. oats, 280,000lb. hay and 375 buckets wine in one day alone. Exactly what requisitions were made during the subsequent period we do not know, but an indication of their magnitude can perhaps be found in the fact that the demand for wine alone rose to 677 buckets per day from 26 November onward. To enjoy these replenishments, the *Grande Armée* was granted three days' rest even though this meant that Kutuzov was allowed to slip away at Hollabrünn.

Our sources have left us with scant details concerning how the army fared on its way to Austerlitz. On 20 November, Murat informed Napoleon that the supplies for 300,000 rations had been found at Pressburg, and this was followed by an order to establish ovens capable of producing 60,000 rations per day at Spielberg. It is also useful to note that the *Grande Armée* now stood with its back to Bohemia, a country which at this time was classified by experts as sufficiently rich to support an army,[29] and that the distance to Vienna was by no means excessive even for horse-drawn vehicles.

Operations having come to a standstill, Napoleon's army soon found itself in supply difficulties even though requisitions were being made far and wide.[30] Luckily for him, the allies were faring still worse and were ultimately compelled to attack the *Grande Armée* on pain of disintegration.[31] As a result, a battle was fought; and the *Blitzkrieg* from Boulogne to Austerlitz was over.

Many roads to Moscow

As sole commander in chief possessing absolute authority Napoleon was not in the habit of drawing up detailed memoranda, hence there is no information as to how he assessed the performance of his army's supply apparatus nor what lessons he drew from it. To judge from his subsequent actions, however, it appears that the Emperor was at first well-content with his own system. When the *Grande Armée* took the field again the following year it was still operating in exactly the same way and, indeed, achieved similar results by thrashing the Prussians in a war that lasted just six weeks. This time the army took with it provisions for ten days, and prior to the battle of Jena–Auerstädt lived largely off the country. After the victory, huge requisitions were made at such towns as Weimar, Erfurt, Leipzig and Küstrin, enabling the troops literally to wallow in luxury during the months October–December 1806. With the new year, however, the army entered Poland where very little could be found, and it became necessary to establish a regular line of communication back to Saxony. It was at this stage that Napoleon first came face to face with considerable opposition and partisan activity between the Oder and the Vistula. The transport service was organized by Daru with the help of German contractors. Use was also made of the waterways (Havel, Spree, Oder, Wartha, Netze, Bromberg Canal, Vistula) which, since the winter was exceptionally mild, were serviceable from the middle of February onward. None of these measures fully met requirements, however, and so the establishment of a military train consisting of seven transport-battalions with 600 vehicles each was decreed on 26 March.

Again, we have little information on the logistic aspects of the 1809 campaign. This time the Austrians stole a march on Napoleon and took him by surprise, so that there was no time to form a proper base even if he had wanted to. Nevertheless, some stores were apparently concentrated at Ulm and Donauwörth,

and following his experiences of 1805, the Emperor organized a
flotilla of boats to forward these stores along the Danube. Since,
however, the campaign unfolded with extreme rapidity from the
start of the French advance on 17 April to the occupation of
Vienna just three weeks later, it is unlikely that this measure was
of much use. Rather, the troops must have lived by drawing on
the reserves carried along – this time put at twelve days – and off
the country.[32]

Looking back upon these campaigns, all so successful in spite
of being conducted – involuntarily, for the most part – on a logistic shoestring, it seems ironical that Napoleon's first major failure
resulted from the operation that he had most carefully prepared.
For the invasion of Russia was no ill-conceived adventure as it is
so often described. Of all Napoleon's wars, it was the one for
which he had assembled means, human and material, out of all
proportion to anything that went before, not merely in his own
day but during the previous centuries as well.

It is inconceivable that Napoleon did not know what was clear
to every other military man of the time, namely that in 'the waste-
lands of the Ukraine' (Guibert) it was impossible to live off the
country. He had in fact written his son in law that 'the war in
Poland will hardly resemble that in Austria; without adequate
transportation, everything will be useless'. Not only had the
Depôt de Guerre received, as early as April 1811, an order to
amass all possible information on Russia, but he was familiar
with the history of Charles XII's Russian campaign and must
have known that the Swede was faced not merely with a thinly-
populated country but one that had been systematically ravaged
by the retreating enemy.[33] Nor were his experiences during the
1809 campaign – the last he had commanded in person – of a
kind to encourage logistic neglect. After the halt inflicted on it at
Aspern, the *Grande Armée* had found itself cooped up in the
island of Löbau and laboured under great supply difficulties.
Hence preparations for a defensive war against Russia had
started long before the Emperor made up his mind to attack.
For example, 1,000,000 biscuit rations were ordered for Stettin
and Küstrin as early as April 1811.[34] At the same time, Napoleon
also increased the size of his train service. These preparations,
however, had a purely precautionary character – proof that
Napoleon was reckoning with the eventuality of a Russian attack
and did not intend to be caught off-balance.

Towards the end of 1811, the measures taken to improve the army's logistic system in Poland began to assume a more offensive character. In January 1812 the provisioning of Danzig with victuals was ordered. By 1 March, subsistence to last 400,000 men and 50,000 horses for fifty days was to be concentrated there. In addition, further 'large stores' were to be stockpiled on the Oder.[35] To carry these provisions, the train service was vastly expanded until it numbered no less than twenty-six battalions (in proportion to the army's size, rather more than those accompanying Moltke's 'modern' forces in 1870), eight of which were equipped with 600 light and medium vehicles each, and the rest with 252 four-horse wagons capable of carrying 1.5 tons.; 6,000 spare horses were also made available.[36] The decision to concentrate on heavy vehicles has often been criticized, for they proved unable to negotiate the atrocious Russian tracks. More than most modern writers, however, Napoleon was aware that lighter carts would have spelt an even greater number of horses and, consequently, increased difficulty in feeding them.[37]

Similarly, the equipment of the army with ammunition was carried out on a grandiose scale. Here, the main depôt was at Magdeburg whence huge quantities of shot and powder were shipped down the Elbe and so on to East Prussia.[38] A note of 1 May 1812 gave the stocks available at Danzig, Glogau, Küstrin, Stettin and Magdeburg as follows:[39]

for	pounders	rounds
59	24	82,612
34	20	32,804
330	12	226,568
69	8	53,835
314	6	365,982

All of this was in addition to the siege artillery.[40] Thus there were available between 670 and 1,100 rounds per barrel for most calibres, figures that do not compare at all badly with those of an industrialized and highly militaristic Germany one hundred years later.

Exactly what Napoleon intended to do with all these preparations is hard to say, for none of his operational plans – if such ever existed – have survived. It can only be conjectured, therefore, that he recognized that no horsedrawn supply system, however well organized, could sustain him all the way from the Niemen to Moscow, as the following figures will show. Even if he were to arrive in the Russian capital with only one third of his original 600,000 men, taking sixty days to do so (in fact he took eighty-two) overall consumption during this period would have come to 18,000 tons for the men alone, which was almost double the overall capacity of his supply trains, which, moreover, would have to supply other parts of the army as well. Furthermore, daily consumption in Moscow would have come to 300 tons, and to cater for this at a distance of 600 miles from base (assuming a very high performance of twenty miles per day by the supply columns) 18,000 tons of transport would have been needed. Regardless of whether he used his trains as rolling magazines or to shuttle (in relays) between the army and the frontier, therefore, there was not the slightest chance of feeding the troops in this way during an advance to Moscow.

As it was, Napoleon took with him into Russia twenty-four days' provisions, of which twenty were transported by the train battalion and 4 on the men's backs. However, it is improbable that he expected the campaign to be over in twelve days (on the assumption that the army would have had to march back as well), as one modern writer[41] has implied. Rather, he must have been thinking in terms of a war lasting for about three weeks (this, we remember, had been the duration of both previous campaigns in the Danube Valley) during which time he would have penetrated Russia to a depth of over 200 miles, hopefully more than enough to catch the Tsar's army and bring it to battle. After that, provisions would have been supplied to the victors by the vanquished, as was Napoleon's normal practice.[42]

Whatever the Emperor's exact intentions, there can be no doubt that he allowed logistic considerations to play a crucial part in the planning of the campaign. The supply from base of fodder for the 250,000 horses accompanying the army being an utterly insoluble problem, the beginning of the war had to be postponed to the end of June. He also started off from Kovno and proceeded to Vilna for logistic reasons, for whereas a deployment further north would have met very great obstacles in view of the appalling Polish roads (which Napoleon knew from his experiences in 1806–7)[43] one further south would have made it difficult to use the Niemen in order to supply the army. Whether, during the last weeks before the start of operations, the Emperor intended to break through the enemy's centre, or to envelop him from the north, or the south, we do not really know. Nor does it matter, for logistics determined his strategy in a way that would have delighted Louvois.

Similarly, the Russians' plans for the defence of their country also rested on logistic considerations. That only the factors of distance, climate and supply could defeat the French army – the largest ever assembled, and commanded as it was by one of the greatest generals of all time – the Tsar's advisers unanimously agreed. The question was not whether to retreat, but where and how far. Here political considerations seem to have played some role, for the nobles were afraid that too long a retreat would lead to a revolt of the serfs.[44] Another difficulty facing the Russian planners was how to compel the French to follow their retreating forces instead of ignoring them. To solve both problems, the Tsar's chief military adviser, General Pfuel, set up a fortified camp at Drissa, midway between the roads to Moscow and St Petersburg, assuming that Napoleon would not be able to bypass the camp during his advance against either city. If the French were to follow the Russians to Drissa, they would find themselves operating in poor country where they would find only a fraction of the necessary supplies.[45] At the same time, another Russian army was to operate in Napoleon's rear, making the task of feeding the French troops more difficult still. The interest of this much maligned plan does not lie in its supposed weaknesses, ridiculed as these have been by Clausewitz. It lies instead in the fact that, like Napoleon's own schemes, it rested on logistic rather than strategic considerations.

In the event, the plans of both sides came to nothing. Napoleon crossed the Niemen on 23 June. Two days later, he was already angrily exhorting Berthier to send provisions to Tilsit where the army 'is standing in great need of them'.[46] Such cries of distress were to become a standard feature of the campaign, and there is no need to go through them all. The main reasons for the failure of Napoleon's logistic plans were as follows. First, the army's supply vehicles proved too heavy for the Russian 'roads', a problem aggravated still further when thunderstorms during the first fortnight of the campaign turned them into bottomless quagmires.[47] Secondly, the river Vilnya, on which Napoleon had relied for shipping supplies to Vilna, turned out to be too shallow to allow the barges through. Thirdly, discipline in the army was lax, with the result that the troops plundered indiscriminately instead of carrying out orderly requisitions, the outcome being, paradoxically, that the officers – at any rate, those who refused to take part in such excesses – starved even when the men found enough to eat.[48] Furthermore, the troops' indiscipline caused the inhabitants to flee, and made the establishment of a regular administration in the army's rear impossible. Fourthly, some of the troops, notably the German ones, simply did not know how to help themselves. Finally, there was deliberate destruction by the Russians. This sometimes assumed disastrous proportions, e.g. early in July when Murat reported he was operating in 'very rich country' which, however, had been thoroughly plundered by the Tsar's soldiers.

Though innumerable grumblings and complaints accompanied the *Grande Armée*'s march to Moscow, this did not mean that all the troops were invariably and equally badly off. In particular, the advance guard usually fared better than the rest of the army because it was the first to enter new territory. The rearguard – consisting of Napoleon's darlings, the Imperial Guard – also did comparatively well, either because the Emperor took good care of them or, which seems more likely, because they were marching at some distance behind the rest of the army and thus found the inhabitants of the villages on the way already returning to their dwellings.[49] Davout's corps, marching well to the south of the main force in an attempt to cut off Bagration's 2. Russian Army, repeatedly found 'more [food and fodder] than I ever hoped', and at one point deemed it necessary to emphasize that, in spite of the glowing reports, the troops were not wallowing in luxury.[50] On the other hand, the forces operating away from the *Grande Armée*'s main body – those under Princes Poniatowsky and Jerome in particular – fared worse than most, as did Murat's cavalry, who found it so difficult to obtain fodder that half of their horses had died by the time the Dvina was reached.

At that point, however, the thinly populated areas of Lithuania and Belorussia had been left behind, and the worst of the army's suffering was over. Napoleon must have known before he embarked on the campaign that the regions round Smolensk and Moscow were comparatively rich – the number of inhabitants varied from seventy to 120 per square mile.[51] This, in all probability, was the major reason behind his decision to continue eastward after the disappointment of his hopes to beat the Russians near the frontier. Nor did the Emperor's calculations prove wrong. From about the middle of July onward, unit after unit reported that while the villages on the way had frequently been pillaged the country was 'improving with every step forward', 'very good and well cultivated', 'magnificent', 'covered with wonderful crops', and 'offered the most abundant harvest'.[52] That many of these reports originated with the Guard, which was making up the rear and could therefore expect to find the country it entered already plundered and devoid of resources, proves conclusively that the *Grande Armée*'s troubles stemmed less from any shortage of provisions in the country than from a lack of discipline that not only drove away the inhabitants but also extended to pillaging the army's own convoys. Nevertheless, even those eternal grumblers, Eugène and Schwartzenberg, found the going easier.[53] As a private letter sent back from around Smolensk on 22 August put it, 'the country which we are entering is very good; the harvest is abundant, the climate agreeable. You may imagine that it offers great resources. . .the health of the army is excellent. We do not lack either bread or meat. As for wine, there isn't as much as in Bourgogne, but we have no reason to complain.'[54]

The present chapter is not intended to belittle the logistic difficulties facing Napoleon's invasion of Russia, especially in view of its disastrous outcome. It should be recognized, however, that the worst shortages were experienced during the first two weeks of the advance (i.e. precisely the period for which Napoleon had made his most careful and extensive preparations) and that the situation gradually improved afterwards. Also, the *Grande Armée*'s problems were at all times – including the retreat from Moscow[55] – largely due to bad discipline. This, of course, was itself partly due to logistic shortages. However, the fact remains that those units whose commanders were strict disciplinarians (e.g. Davout's) consistently did better than the rest, while the Guard even managed to keep such good order that, far from running away, the inhabitants enthusiastically welcomed it. Nor is it true, as is so often maintained, that the country as a whole was too poor to support an army. Writing from Drissa early in July, Murat – operating as he was in an area which Pfuel had selected for the erection of his fortified camp precisely because it was supposed to be without resources – informed Napoleon that while the region around was tolerably well provided it would be possible to exploit it only after a proper administration was set up and an end put to the troops' marauding.[56] All our sources agree that the country became more abundant the closer one got to Moscow, which, together with the (justified) hope that the Russians would not give up their capital and holy city without a fight, was probably why Napoleon went there instead of terminating the campaign at Vitebsk.

The above facts also dispose of the attempt, recently made,[57] to refute Clausewitz's objections to Pfuel's plans for the defence of Russia and show that the latter were correct after all. That Pfuel relied on the logistic wisdom of the age is undeniable, but this in itself is not enough to prove him right. The plan's main fault was that the camp at Drissa was too close to the frontier, and would surely have been reached by the *Grande Armée* regardless of the economic circumstances on the way; that more than a few days' starvation were needed to put France's troops off the scent was demonstrated time after time throughout the wars of the Revolution and the Empire. As it was, Napoleon's army in 1812 succeeded in reaching Borodino in sufficiently good order to beat both the Russian armies combined, despite the enormous distances and logistic problems involved. What would have been the fate of Barclay de Tolly's troops had they waited for the Emperor behind their trenches at Drissa is only too easy to imagine.

While bad discipline did, as we saw, play a crucial role in the failure of the campaign, the arguments attacking the French army's technical proficiency appear largely unfounded. It may be true that Napoleon reduced the supply of his army over the 200 miles closest to the frontier to an arithmetical problem, not taking sufficient account of the ubiquitous 'friction' of war, though it must be remembered that he had to make his calculations for the campaign on the basis of inadequate information. That the quality of the French train-personnel left something to be desired is certainly shown by the surviving evidence, but then this was due above all to the difficult circumstances[58] and not to any lack of experience, for the train-organization had by this time been in existence for five years. Above all, the claim that the French troops and commanders did not know how to live off the country is utterly ridiculous and unworthy of its author; their expertise in this field was deservedly famous and indeed had reached such heights that it helped them, within the unprecedently brief period from 1800 to 1809, to overrun all Europe and set up an Empire the like of which the world had never seen.

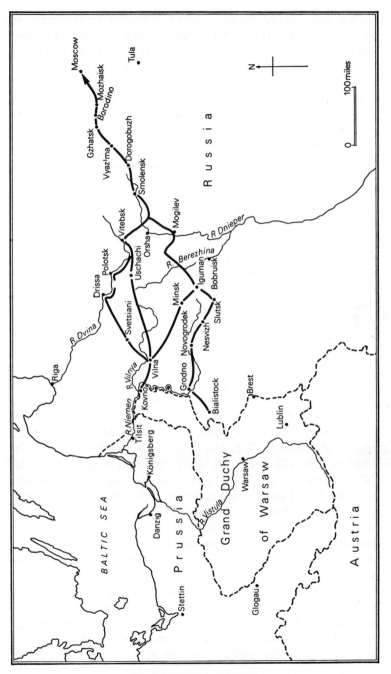

4. The Campaign of 1812

That the *Grande Armée* suffered enormous losses during its
march to Moscow is true,[59] as is the fact that hunger and its consequences – desertion and disease – played a large part in causing
these losses. It would, however, be unwise to attribute this solely
to the problems of supply. The need to protect enormously long
lines of communication and to leave garrisons behind, and the
effect of distance *per se* were also factors of major importance. As
regards the army's material losses, there is reason to believe that
much if not most of the equipment abandoned on the way to
Moscow was later retrieved.[60] In 1812 Napoleon's main force
marched 600 miles, fought two major battles (at Smolensk and
at Borodino) on the way, and still had a third of their number
left when entering Moscow. In 1870, as in 1914, the Germans,
operating over incomparably smaller distances, in very rich
country and supported by a supply organization that became the
model for all subsequent conquerors, reached Paris and the
Marne respectively with only about half of their effectiveness.
Compared with these performances, excellent as they were, the
French Army of 1812, for all its supposedly worthless service of
supply, did not do too badly.

Conclusions

In reading modern accounts of Napoleon's logistic system, one
comes across so many misunderstandings that the question as to
their origins inevitably presents itself. That only a first class writer
could have given rise to these errors is beyond question. The
threads of investigation in fact point to the greatest military critic of all time, Clausewitz. It is not altogether surprising that he
should be the source of these errors, for the Prussian's whole doctrine was based on the assumption that Napoleonic warfare was
qualitatively different and represented an entirely new departure. It was, after all, Clausewitz who called the Emperor
'the God of War', and who invented the term 'absolute war' in
order to describe what he believed to be the essence of his
system.

Clausewitz assumed as a matter of course that a revolution in
warfare as fundamental as that effected – in his view – by Napoleon
could not have taken place without an equally profound change
in the logistic methods employed, and this led him to invent
an army which did without magazines, lived off the country,

paid no attention to considerations of supply and sometimes seemed to grow wings in its marches from one European capital to another. That this picture was exaggerated, contemporaries were well aware. Writing about the campaign of 1805, the much-maligned Bülow rightly observed that the French army had never been able to do entirely without magazines and attributed its speed of movement less to any freedom from logistic shackles than to the absence of heavy baggage.[61] If an excuse for Clausewitz can nevertheless be found in the fact that he was so close to the events and thus lacked perspective, this does not apply to modern historians, especially as they have shown Napoleonic warfare – his strategy, tactics, organization etc. – to be the logical outcome of progressive developments originating in the previous thirty or forty years.[62] That the *only* aspect of Napoleon's operations which, to this day, is believed to have been not merely completely original but actually regressive in comparison with earlier practice is remarkable and should in itself have led to second thoughts.

In this chapter, we have deliberately concentrated on the two campaigns representing the extremes of Napoleonic warfare – the unprecedentedly successful war of 1805 on one hand, the disastrous one of 1812 on the other. We saw that, during the former, the French managed without either magazines or a well-organized transport service because there was insufficient time to set them up. We saw also that the number of vehicles with which Napoleon originally intended to provide his army was exactly similar, in proportion to its numerical strength, to that accompanying the forces of the much maligned 'unfortunate General Mack'. The fact that he did not succeed in getting hold of this quantity of wagons or in manufacturing the provisions necessary to fill them compelled Napoleon to change his plans and shift the direction of his advance from a poor, thinly-inhabited part of Germany to another that was richer and offered more resources. In this respect, far from being free from the tyranny of logistics, Napoleon's practice was similar to that of Wallenstein and Gustavus Adolphus 170 years previously.

Having arrived in Ulm, however, the Emperor realized that he would not be able to go on in the same way. He therefore organized a transport service on an unprecedentedly large scale, and his provisions for its establishment may well be regarded as a model of their kind. He caused huge magazines to be set up in

the towns of Bavaria and provided for convoys of wagons and
boats to bring them to the advancing army. If these arrangements
failed to have much effect on the conduct of operations, this was
due above all to the intolerable crowding of such enormous forces
on a very small number of roads. At one stage as we saw, no less
than five corps had to share a single road between them. Under
such conditions any army was doomed to suffer logistic hardships. Even present-day armies with their tens of thousands of
mechanical vehicles[63] cannot easily solve the problem of supplying such a dense mass of men. If some supplies nevertheless got
through, if the army did not starve, nor the majority of its units
disintegrate, this was due, not to any supposed neglect of logistic
considerations but to a triumph of foresight, organization and
leadership. Such a triumph, however, could only be achieved by
a genius of Napoleon's own calibre, and it is therefore no accident that, precisely during this period, we find him dozens of
miles behind his advanced spearheads and allowing his care for
logistic organization to seriously interfere with his conduct of
operations.

It is scarcely surprising that Napoleon should have rested content with the military machine – its administrative and logistic services included – that had made a victory of the magnitude of
Austerlitz possible, but he nevertheless took sufficient cognizance
of its shortcomings to order that larger reserves should be carried
along during the subsequent campaigns. In 1807, moreover, he
took a step that was, at the time, revolutionary: for the first time
ever, the *Grande Armée* was given, in addition to the vehicles
accompanying the troops, a regular train service. This consisted,
no longer of requisitioned and hired vehicles and drivers but of
fully militarized personnel and equipment. Far from reverting to a
more primitive practice, therefore, Napoleon – in this field as in
most others – was ahead of his rivals; that the train service, novel
as it was, did not at first fulfil all expectations is not astonishing.

This brings us to the campaign of 1812. As we saw, the invasion of Russia was not started without adequate preparations; on
the contrary, Napoleon's measures exceeded anything Louvois
ever dreamt of. Even so, the technical means of the age made it
hopeless even to try and feed the men – much less, the horses –
from base, a fact of which Napoleon was fully aware and which
led him to plan a campaign which, for the most part anyway,
would be won and finished *before* the defects of his logistic appara-

tus would become apparent. In the event, the blunders of his subordinates – above all his brother, Jerome – made it impossible to
cut off and annihilate even that part of the Russian army which
seemed within his grasp, with the result that he found himself
at Vitebsk – the furthest point to which his logistic system
might, albeit only by means of a supreme effort, have proved
adequate – without having achieved his objective. Faced with the
alternative of either retreating or making another attempt to
force the enemy to battle, the Emperor hesitated, vacillated, and
finally decided in favour of the latter. This was made easier by
the fact that the poorest part of Russia had already been left
behind and that requisitions could be made in a country that
became richer the further east he went. That good results could
be obtained provided discipline was kept is shown by the fact
that the Guard reached the Russian capital virtually intact.

During the advance to Moscow, the train service of the *Grande
Armée* broke down as Napoleon always expected it would. However serious its own shortcomings, there can be no question that
this was a consequence of the impossible circumstances, and had
comparatively little to do with the inexperience, indifference or
corruption of the personnel. In 1870, and again in 1914, the Germans with their immeasurably superior and excellently organized
supply apparatus utterly failed to feed their armies from base and
had to resort to requisitions, in spite of the fact that the distances
involved were smaller by far, the roads available very much
better and more numerous.

Coming now to the actual methods of requisitioning, Napoleon
had at his disposal an unrivalled administrative machine in the
shape of the *ordonnateurs* and the *commissionaires de guerre*, to
whom he owed a large measure of his success. Perfectly aware of
the harmful effect that 'direct' requisitioning had on the morale
and discipline of an army, he tried to avoid such a procedure
whenever possible, either by having supplies collected in advance – as did Marlborough, albeit against ready cash, in 1704 –
or by levying contributions with which supplies were subsequently purchased,[64] as was the standard practice of every
eigtheenth-century commander. Either way, receipts were given
and accounts kept, even in enemy country, the intention being to
settle with the – hopefully defeated – opponent after the war.
Only in extreme necessity, e.g. during the time when 150,000 men
were massed around Ulm in 1805, did Napoleon resort to 'direct'

requisitioning, and this was, of course, suspended again as soon as was practicable and opportune.

Given this dependence on the country it is not surprising that Napoleon's forces, like their eighteenth-century predecessors, ran into logistic trouble whenever they stayed for too long in any given place. This was the case round Mantua in 1796, during the enforced pause before the battle of Austerlitz in 1805, when the army was confined in Löbau in 1809, and during the stay at Moscow. However, perhaps the most revolutionary aspect of Napoleon's system of warfare was precisely that he usually knew how to prevent such pauses from taking place, that he was able to go straight on from strategic march to battle and then to pursuit, and that he avoided sieges. How right he was to do this is also demonstrated by the experiences of his marshals in Spain, where geographical circumstances made siege-warfare inevitable and where one French army after the other accordingly suffered from starvation.

There were many factors that gave rise to the unprecedented momentum that enabled the French armies to do what their predecessors had normally failed to do, namely to march from one end of Europe to the other, destroying everything in their way. These included the *corps d'armée* system which, by dispersing the army's units, made it easier to feed them from the country; the absence of baggage (more important, this, in hampering the movements of eighteenth-century armies than any supposed dependence on supply from magazines); the existence of a regular apparatus responsible for making requisitions; the fact that Europe was now more densely populated than previously (this is stressed, and rightly so, by Geza Perjes); and, to quote Napoleon's own explanation, the sheer size of the French armies which made it possible to bypass fortresses instead of stopping to besiege them. In the final account however, none of these 'material' factors adequately accounts for Napoleon's success. This suggests that, even in a study so mundane as the present, the role of genius should not be underestimated.

3

When demigods rode rails

Supply from Napoleon to Moltke

Scarcely had the guns of Waterloo fallen silent than soldiers everywhere began the process of studying and analysing Napoleon's campaigns with an eye to learning from them lessons for the future. Since the mobility of the French armies throughout the revolutionary period was legendary and universally recognized as a crucial factor in their success, the problem of logistics was subjected to a particularly searching examination, a process, indeed, which had started several years previously as commanders strove to imitate Napoleon's methods *ad hoc*. Thus, for example, the establishment of wagons, pack horses and baggage authorized to Austrian formations was drastically cut down as early as 1799–1800, and again after the defeats of 1805, with the result that, in the campaign of 1809, the Austrians were able to match the marching performances of the French for the first time.[1] Again, on their way from Saxony to the Rhine in 1813, the allies showed that they had absorbed some of the lessons taught by the great Corsican.

Following this attempt to study and emulate Napoleonic logistics it did not take long for two opposed schools to appear. Of these, one is perhaps best represented by André de Roginat, a French officer who had taken part in the Emperor's wars and who published his reflections on them as early as 1816.[2] In a chapter entitled 'Des grandes operations de la guerre offensive en Europe', Roginat subjected Napoleon's administrative arrangements to a scathing criticism and reached the conclusion that his ultimate failure was due above all to the inadequate attention he paid to the lines of communication. In Roginat's view, strategic penetration deep into the enemy's country was all very well as long as it was carried out by small armies. Modern forces, however,

were enormous in size and made demands in subsistence, ammunition and replacements on a scale that was more enormous
still. Roginat further emphasized the difficulties that inevitably
follow an army living off the country, including desertion, indiscipline, and trouble with the population. Calling the Austerlitz
campaign 'the height of madness', he accused Napoleon of allowing 300,000 men to starve to death in Russia and another 200,000
in Saxony. To remedy all these evils, Roginat suggested that the
future lay in a 'methodical', step by step, system of warfare.
Armies were to be loaded with a maximum of eight days' provisions, then to advance to a distance of no more than thirty to
forty leagues from their base, at which point they were to stop,
take stock, and wait for the *armée de reserve*, which Roginat considered essential for any successful military enterprise, to catch
up. Stores would then be accumulated and a new base established. Not until everything was complete could the process be
repeated.

Very different were the deductions of another, and greater,
writer, Karl von Clausewitz. When it came to nuts and bolts
Clausewitz was, surprisingly enough, less impressed with the
rapidity of the Emperor's strategic movements than were the
majority of his contemporaries. Pointing out that the need to
make requisitions, especially if it could not be done directly by
the troops, could impose delays quite as bad as those made necessary by a system of supply from base, he used Murat's famous
pursuit of the Prussians in 1806 in order to demonstrate that there
was nothing here that Frederick II, for all his great train and
baggage, had not been able to do equally well.[3] Clausewitz, however, thought that the French system – or lack of it – became
more advantageous in proportion as distances grew; it alone had
made possible those tremendous marches from the Tagus to the
Niemen. Whatever his reservations, therefore, Clausewitz was
ready to see the wave of the future in supply by the country.
Contradicting his own detailed examination, he concluded that
such was the superiority of a war carried on by means of requisitions over one dependent on magazines that 'the latter does not
at all look like the same instrument'.

While this may be taken as the considered *theoretical* opinion of
the post-Napoleonic generation, attempts to put it into practice
soon showed that mere plunder was an unsatisfactory method of
feeding an army. Thus, the Russian Administrative Regulations of

1812 charged the intendant general to use 'requisitions, purchase and contracts to exploit the resources of occupied countries for the army's benefit, while only resorting to our own stocks. . .in special cases'. In practice, however, Russian commanders did not possess enough initiative to maintain their men in this way, with the result that 'much suffering' took place during the campaigns of 1828–9 (against Turkey) and 1831 (in Poland). Consequently, 'ambulant magazines' were introduced in 1846 together with trains of field bakeries and butcheries. These arrangements did not, however, prove themselves during the Crimean War, with the result that innumerable carcasses of men and horses lined the routes to Bulgaria and Sebastopol.[4]

Unlike the Russians, who were forced by geography to conduct their wars in poor and thinly populated areas, the Austrians in 1859 found themselves operating in the rich country of north Italy whose resources should easily have fed the army. Requisitioning was therefore resorted to as a matter of course. The troops' organic transport, however, did not arrive on time, while attempts to organize columns of local vehicles took so long that the troops, who were strictly forbidden to help themselves, starved. In addition, cooperation between the intendant general and the corps was not without friction; so that the entire organization was subsequently characterized as a 'complete failure'.[5]

As organized in 1814–15, the Prussian train-apparatus was a comparatively sketchy affair, consisting of provisions-columns (theoretically capable of carrying four days' food for the army), field bakeries, remount depôts, and ambulance wagons. A Royal order of 1831 organized these into 'train companies', of which one was allocated to each corps, a train company consisting of seven (subsequently reduced to five) provisions-columns, a field bakery column and a 'flying' remount depôt. To manage the service of supply, a body of train officers (with personnel detached from the cavalry) was established in 1816. In the same year, the corps were instructed to appoint commanders of the trains, though the latter were made subordinate to the inspector of trains and thus independent of the formations they were supposed to help maintain.

Apart from the designation of officers who, in time of war, were to command the trains, an organization responsible for operating the service of supply did not exist. Only part of the *materiel* for the trains was stored in depôts and was thus immediately available for use. It was intended to supplement this

with requisitioned horses and vehicles. Similarly, nuclei of qualified NCOs to train the necessary personnel were not established,
and the entire organization had to be built up from scratch during
mobilization.

During the events of 1848 and 1849, these arrangements were
tested for the first time and, as might have been expected, found
completely inadequate. The main problem lay in the quality and
quantity of the personnel. The commander of the trains, a Major
von Freudenthal, was 69 years old while his principal collabora-
tors were veterans of the campaigns of 1812 and 1813, the
youngest of them being aged 55. No officers were available to
command the field hospitals and bakeries, which consequently
had to be entrusted to NCOs, some of them invalids who had never
previously even ridden a horse. Equipment, especially clothing,
was so short that some of the men spent fourteen days at the
depôts before finally being issued with their uniforms. Arrangements for feeding the horses had not been made, and dreadful
confusion ensued when officers tried to group men whom they did
not know into units that did not exist. Under these circumstances,
the trains were unable to reach the area of deployment in time,
which led that soldier-prince, Wilhelm of Prussia (subsequently
King and then first German Emperor) to write that supply and
transportation were the weakest points in the entire Prussian
army organization.[6]

In 1853, thoroughgoing reforms were accordingly initiated by
the Minister of War, von Bonin. In January, the training in peacetime of train officers and NCOs was provided for; this was followed three months later by a more detailed order signed by
Frederick Wilhelm himself, specifying that each corps should
detail a staff officer for the establishment of a 'nucleus' (*Stamm*)
of train personnel, and that this personnel should engage in sup-
ply-exercises during fourteen days each year. In 1856, these
'nuclei' of train personnel were expanded into regular train battalions which, moreover, were now made subordinate directly to
their respective corps headquarters. A train battalion consisted of
a staff, five provisions-columns (overall capacity 3,000 *Zentner*
flour, approximately eight days' consumption), a field bakery, a
remount depôt, one main hospital and four field hospitals. However, when the army was mobilized against France in 1859 many
of these turned out to exist on paper only. After demobilization
measures aimed at bringing the trains of all corps up to establish-

ment were introduced. Finally, in June 1860, there came yet another reorganization which turned the train troops into an independent arm (*Waffe*) of the army and gave them their own inspector-general. This inspector was subordinated, not to the General Staff, but to the Ministry of War, and was responsible for training and nominating the train-commanders. Each of the nine corps into which the Prussian army was now divided was given its own train battalion which, in peacetime, consisted of 292 officers, NCOs and men. The personnel were listed as combatants, and accordingly wore the regular army uniform.

Complete in theory, these arrangements were not really tested in the Danish campaign of 1864 because Prussia put into the field on that occasion only 43,500 men (rather less than two complete corps), 12,000 horses and 100 guns – a fraction of her overall military strength. During the period of concentration, the army as a whole was provisioned by quartering. However, magazines of flour and fodder were set up around Kiel, as was a reserve wagon-park, 1,000 strong. It being winter, some difficulties were experienced with the weather because the roads were iced over, which made marching difficult. As distances were small and the country rich, the organization functioned smoothly on the whole, though problems were caused by the lack of training of some of the personnel. In any case, it does not appear that the experiences of this little war led to any important changes in the service of supply.[7]

Very different were the experiences of the Prussian–Austrian War of 1866. Though the train organization should theoretically have been capable of catering to the army's needs, the sheer size of the problem was overwhelming; 280,000 men, concentrated in a single theatre of operations, were to be supplied from base, an enterprise far exceeding anything previously attempted, with the exception of Napoleon's ill-fated Russian adventure. As in 1864, the troops were fed during the period of concentration mainly by quartering, supplemented by free purchase. It was intended, however, to supply the advance against Austria from the rear, 1. Army in particular making its arrangements as if Saxony, where it was to operate, were a desert.[8] In the event, nothing came of these plans. The trains generally succeeded in keeping up with the troops until about 29 June (a time, however, when the invasion of the Austrian territories had not yet got very far) but were subsequently left behind and did not succeed in catching up until *after* the battle of Königgrätz was fought and won.[9] While the trains became entangled in monumental traffic jams and battled for priority on the roads, the troops were fed by quartering, requisitioning, and sometimes not at all. As Moltke himself wrote to his Army commanders on 8 July, this failure was caused by the following 'abuses':

      a. The trains were crowded off the roads by columns of infantry, cavalry and artillery, sometimes being immobilized for days on end and thus became separated from the troops they were supposed to supply.

      b. Field police, who should have been available to supervise marching discipline, were frequently employed on other jobs. Moreover, their commanding officers used the fact as an excuse for not carrying out their proper duties.

      c. Supply trains tended to become inflated by unauthorized vehicles, many of which were not suitable to military purposes.

      d. Congestion was especially frequent in defiles and other narrow places, because of the lack of leadership. Columns and individual vehicles behaved as they pleased, often resting on the roads and thus blocking them.[10]

After Königgrätz, the army continued to live mainly by requisition, which compelled Moltke to suspend the standing orders and permit corps, divisions and even battalions to skip the services of the quartermasters and look after their own supplies in order to save time.[11] However, Bohemia did not yield very much; villages were frequently deserted, and all transport had been taken away by the retreating Austrians. The only item that was really plentiful was meat, while bread was very short, sometimes for weeks on end. This occasionally affected operations, e.g. when 2. Guards Infantry Division was brought to a halt on 19 July.[12] It is the opinion of at least one expert that, had the campaign been prolonged, these shortages might have been catastrophic.[13] Even as it was, the strain of weeks of marching coupled with undernourishment led to a bad outbreak of cholera. Fortunately, however, the Seven Weeks' War ended a mere twenty days after Königgrätz, before the shortcomings of the logistic apparatus were even properly understood.

Though the Prussian army of 1866 did have a well organized supply apparatus, it was not, in regard to the actual methods used to feed the troops in the field, very much more modern than Napoleon's *Grande Armée* sixty years previously. The same was true for the supply of infantry ammunition. In spite of being issued with the needle gun, Moltke's soldiers were able to carry all their ammunition inside the corps, a total of 163 rounds per rifle being distributed between the regimental wagons, the battalions' carts and the men's backs. Arrangements to provide for a constant flow of ammunition from the rear did not exist, nor were they needed in view of the extremely modest consumption. Throughout the campaign, no more than 1.4 million rounds were expended, an average of seven per combatant.[14] As a result, the number of rounds transported in the various wagons was reduced after 1866, while that loaded on the infantryman's back was increased until it formed one half of the total supply – another indication of the still comparatively primitive state of the logistic services.[15]

One very significant consequence of the attempt, however unsuccessful, to feed the troops from base, was a new and severe limitation on the maximum number of troops that could be marched over a single road. In Napoleon's day, this limit was imposed by the need to secure for each unit a piece of country sufficiently large to allow it to forage; under Moltke's organization, however, the number was dictated by the distance that his horse-drawn supply trains could travel each day. Assuming that this was 25 miles, the maximum length of a column of marching troops could not be allowed to exceed 12.5 miles if the convoys in the rear were to reach the advance guard and travel back to replenish in a single day. In theory, it was possible to multiply this figure by employing large numbers of wagons divided into echelons each carrying a day's supply, but this would involve the regular shuttling of columns past each other in opposite directions, an operation not at all easy to carry out on the roads of Central Europe in the eighteen sixties. In practice, it was found that no more than one corps – 31,000 men – could be marched over each road, which led Moltke to make the celebrated dictum that the secret of strategy was 'to march separately, fight jointly'. As it was, it was not always possible to observe this rule. After Königgrätz in particular, the shortage of victuals at 1. Army stemmed partly from the fact that its three corps were crowded on a single road so that the supply trains could not get through.[16] Even so, recognition of the principle led each of the three Prussian Armies approaching the battlefield along five roads, with good lateral communications between them, whereas the entire Austrian army had to march by two roads only.[17]

A joker in the pack

The second half of the nineteenth century was the great age of the railways, and no part of Moltke's system of warfare has received so much attention and praise as the revolutionary use made of this novel means of transportation for military purposes. Before going on to analyse the role of the railways in the Franco-Prussian War, therefore, it is necessary to say a word about their development as an instrument of war and conquest.

As is well known, one of the first to suggest that armies could benefit from the utilization of railways was Friedrich List, an economist of genius who, in the 1830s, foresaw that a well conceived railway net might enable troops to be shifted rapidly from one point to another hundreds of miles distant, thus multiplying numbers by velocity and enabling them to concentrate, first against one enemy, then against another. Surprisingly enough, the first to grasp the full military potentialities of this were the Russians.[18] In 1846, they moved a corps of 14,500 men, together with all its horses and transport, 200 miles from Hradisch to Cracow in two days by rail. This was followed, four years later, by the Austrians moving 75,000 men from Hungary and Vienna to Bohemia, this being perhaps the first time when the railways played an important part in international power politics, by helping to bring about the Prussian capitulation at Olmütz. Seven years subsequently, however, it was France's turn to give a startled world an object lesson in the strategic use of railways. From 16 April to 15 July, 604,381 men and 129,227 horses were transported by rail, involving all the French lines then in existence, of whom 227,649 and 36,357 respectively went directly to the theatre of operations in Italy.[19]

In Prussia, by contrast, the idea that the railways might be useful for military purposes at first met with nothing but opposition. The heirs of Frederick II echoed his saying that good communications only made a country easier to overrun. Attempts by commercial interests to construct new lines often met with determined opposition on the side of the army, which feared for the safety of its fortresses, and a committee set up to deal with the question concluded, in 1835, that railways would never replace high-roads.[20] This kind of thing went on until 1841, when the debate died down owing to lack of interest.

The Prussian army only began to take a serious interest in railroads during the revolutions of 1848–9, when moving troops by road became unsafe. This, together with the revolutionaries' repeated use of the lines in order to make good their retreat, finally led to a *volte face*. Progress was slow initially, and the Prussian troops using rail in order to get to Olmütz, arrived there in such confusion that they were unable to face a much better organized Austrian force. Though some of the worst shortcomings had been corrected by the time Prussia next mobilized in 1859, the performance of her railways in the military service was still eclipsed by that of France. This was partly due to the fact that there were in Germany at this time dozens of different companies operating lines, between whom there was little coordination and less control. Though efforts to provide some measure of uniformity and central control by the German *Bund* started in 1847, the system that was to permit the unrestricted use for military purposes of the railways in time of war was not completed until 1872.[21]

By the mid-sixties, enough progress had been made to enable the Federal Army – which in practice meant those of Prussia and Austria – to make effective, though scarcely spectacular, use of the railways for military purposes. Accordingly, from 19 to 24 January 1864, the Prussians transported an infantry division (15,500 men, 4,583 horses, 377 vehicles) by rail from Minden to Harburg, using a total of forty-two trains – an average of seven per day – to move this force over 175 miles. Subsequently, use was made of the railways to bring up supplies, an average of two trains per day being employed for this purpose between Altona and Flensburg during the second half of February. Towards the end of this period, it also became possible to send trains further north into Schleswig. Since the scale of these operations was so very small, no lessons of any importance were learnt. Traffic usually worked smoothly, though there was one accident. More significantly for the future, it was found that the unloading of trains created a bottleneck and experiments were accordingly carried out with mobile wooden ramps, unloading trains from the rear instead of sidewards, and the like.[22]

During the campaign of 1866, the railway-network dictated not merely the pace of Prussia's strategic deployment but also its form. In preparation for a war against Austria, it had been the intention of Moltke and the General Staff to deploy the Prussian army around Görlitz, so as to enable it both to cover Silesia and to take in flank an Austrian advance through Saxony to Berlin. The Prussians, however, started mobilizing later than did the Austrians, and to make up for the delay were compelled to make use of all five railways leading to their frontier. This resulted in their forces being deployed on a 200-mile-long arc, and thus Moltke's subsequently celebrated 'strategy of external lines' was born, not because of any profound calculations but as a simple accident dictated by the logistic factors of the time – space and the configuration of Prussia's railway system.

As it was, the use of the railways during the war of 1866 could hardly be regarded as a resounding success. Mobilization, it is true, proceeded smoothly; in twenty-one days, 197,000 men, 55,000 horses and 5,300 vehicles of all kinds were deployed, and it is said that an officer visiting Moltke during this period found him lying on a sofa and reading a book. The subsequent operation of the railways, however, was far less satisfactory. For the first time, it became clear that sending supplies to the railheads was very much easier than getting them from there to the troops. Having failed to allow for supply trains in the mobilization time-tables, Moltke made matters worse by taking his railway expert, von Wartensleben, along to the field. He thereby deprived the entire system of a central directing hand and made it possible for the corps quartermasters to rush forward supplies in great abundance, without taking the slightest notice of the railheads' ability to receive them, the result being that they became congested and then altogether blocked. Thus, towards the end of June, it was estimated that no less than 17,920 tons of supplies were trapped on the lines, unable to move either forward or backward, while hundreds upon hundreds of railway wagons were serving as temporary magazines and could not therefore have been used for the traffic even if the lines had been free to carry them. While bread went stale, fodder rotted and cattle died of malnutrition, field commanders were at least free to ignore the effects of logistics on operations because, as the troops had completely outrun their supply convoys, all connection between them and the railways was lost. Between 23 June, when the first formation crossed the Austrian border, and the end of the battle of Königgrätz, the railways did not, therefore, exercise the slightest influence on the progress of the campaign.[23]

After the victory of Königgrätz, the Prussians found that their inability to make use of the Austrian railways hampered the continuation of the advance into Austria. On 2 July, Moltke demanded that the railway from Dresden to Prague – said to be 'essential...with an eye to our very difficult supply situation' – be opened at the earliest possible moment, but four days later he was forced to recognize that his exhortations were having no effect.[24] In particular, the fortresses of Königstein, Theresienstadt, Josephstadt and Königgrätz blocked the lines to Barduwitz, and inquiries as to whether it would be possible to circumvent them by building emergency lines did not lead to any results before the campaign ended.[25] As it was, the Prussians decided to ignore the fortresses and march towards Vienna, leaving them behind, with the result that, during the second part of the campaign also, the railways were unable to exercise the slightest influence on the course of operations. Meanwhile, the troops lived by requisitioning, impressing what local transport they could find, and behaving as if the railways did not exist, which, effectively, was the case.

In a letter to Bismarck of 6 August 1866, Moltke drew the following conclusions from this.[26] The campaign, he wrote, had demonstrated how easy it was to repair minor damage to the railways: 'the only obstruction of any duration' had been caused by the fortresses, and the chief of the General Staff therefore recommended that Prussia's own railroads should be made to pass through the perimeters of existing ones whenever possible. However, this did not mean that more fortresses should be constructed. The Prussians, after all, had been able to continue their march to Vienna in spite of the blocking of the railways, and at no time had the Austrian fortresses become more than a nuisance in their rear. Consequently it was with rails, not brickwork, that the future lay. What Moltke neglected to say, however, was that, from the completion of the deployment onward, the Prussian army's railways, and to a lesser extent its train service as well, had proved a complete failure and were irrelevant to the outcome of the war.

Whatever part they may have played in the Austrian–Prussian War, Moltke's railways are almost universally supposed to have given him a crushing margin of superiority which, four years later, played a vital role in Germany's defeat of France. A detailed examination of the place of logistics – including supply by

rail – in the Franco–Prussian War will have to wait until later in
this chapter, and here we will only point to the fact that, during
the 1860s, the German – and Prussian – railways were inferior to
the French ones from almost any conceivable point of view. Even
as late as 1868, the anonymous officer who wrote *Die Kriegführung unter Benützung der Eisenbahnen* felt that, the recent
campaign against Austria notwithstanding, the 'overall French
[railway] performance exceeds the Prussian. . .by far'.[27] This was
due to the following factors:

   a. French trains of all kinds travelled faster than German
   ones, this being made possible – in the case of troop
   transports – by an arrangement which required the men to
   take their provisions along, instead of having them
   disentrain in order to be fed at the stations.
   b. Owing to political difficulties, the German railway
   network was less unified, and its material less standardized,
   than the French.
   c. Only twenty-four per cent of the German lines were
   double-tracked, as against sixty per cent of the French
   (this was the case in 1863; after 1866, a start was made to
   improve the situation).
   d. In general, the capacity of French stations was larger
   than that of the German ones; this, of course, governed the
   crucial factor of how long it took to unload.
   e. The quantity of rolling stock per mile of track
   available in France exceeded that of Germany by almost
   one third.
   f. The number of daily trains that could be run over a
   French double line was far larger than its German
   equivalent; in theory, the figures were said to be seventeen
   and twelve respectively, but in practice the French were
   capable, in 1859, of running as many as thirty trains a day.[28]

In the military as opposed to the civilian sphere, the French
advantage was thought to be even greater than these facts indicate, for strategic considerations had guided construction from
the very beginning. This was not the case in Germany, where
political fragmentation meant that economic and local interests
played a much larger role. The French combination of lines running parallel to the frontier, and connecting the major fortresses

with a spider web of routes extending from a central nucleus was considered ideal for the needs of war, comparing most favourably with the 'geometric network' of north–south and east–west lines characterizing Germany's railway system.[29] The assumption that France's railways were superior to Germany's was shared by Moltke himself, and figured large in his intention to stay on the defensive in a war against France.

After the war of 1870, French and German writers accused each other of having constructed their railways with intent to wage aggressive war. That Moltke had a say in the planning of the German network is true. Apart from minor detail, however, he was satisfied that commercial lines were good enough to serve his purpose. On one occasion only did he propose the construction of new lines for purely military ends, but the only response he received to his suggestion was a polite letter of thanks.[30] In 1856, a similar proposal led the Prince of Prussia himself to write to the Minister of War:[31]

> General Groeben. . .advises the construction of railways along the right bank of the Rhine, to which purpose he asks for nine million thalers to be allocated.
>
> The events of recent times have shown that enough private capital is available to build railways, so that burdening the treasury does not appear justified.

Far from driving through military considerations in the construction of new lines, the soldier-prince was perfectly content to leave the development of this crucially important lateral line to private enterprise, though hoping, of course, that the military would benefit later on. At the very least, this seems to show that the claim, so often heard, that the Germans allowed strategic considerations to play a larger role in the planning of their railway system than did the French, requires some evidence to support it.

Finally, it is a curious, but seldom understood, fact that German theoretical writings on the influence of railways on strategy were absolutely wrong. The early exponents – including Ludolf von Camphausen, Moritz von Pritwitz, Heinrich von Rüstow, and one chief of the Prussian General Staff – General von Reyhe – all expected that railways would work in favour of the side operating on internal lines,[32] as did List himself, who hoped that they would 'turn the central situation of the fatherland, which has hitherto been a source of endless evil, into a source of the greatest strength'.[33] In the event, far from facilitating operations on internal lines, the events of 1866 and 1870 were to show that railways helped the belligerent operating on external ones – indeed, in the former case, they compelled him to do so. By contrast, the French railway network in 1870 presented them with magnificent opportunities to exploit the advantages of internal lines, but this did not, of course, prevent them from being thoroughly beaten.

Another error, and one that involved Moltke personally, has been made concerning the effect that railways were supposed to have on the relationship between offence and defence. Here, again, List led the way, writing that:

> The most beautiful thing about it all is the fact that all these advantages [i.e. of operating on internal lines] will benefit the defender almost exclusively, so that it will become ten times easier to operate defensively, and ten times as difficult to operate offensively, than previously.

In List's view, 'greater speed in movement always assists the defender', the reason being that he 'must adapt his moves to that of the attacker'. A well developed railway system would therefore raise the defensive power of a great nation 'to the highest degree available', to the point that, in the view of List and others, war would become altogether impossible and peace reign on earth[34] – another of those predictions that are apt to accompany the appearance of new instruments of war but which somehow seem doomed to be always disappointed.

List was no military man, and derived his conclusions merely from 'a healthy human understanding'. However, his opinion was shared by Moltke, whose reasoning was that whereas a defender would have full use of his own network, the attacker would not be able to rely on any lines in advance of his front.[35] Hence, the railways would help the defence more than they did the attack, a conclusion not substantiated by the fact that Moltke did, after all, wage some of the most successful offensive campaigns in history, which were followed by half a century of attempts to show that railways had made the attack into the best, indeed the only, way to fight a war.

Exactly what one is to make of these facts is not easy to say.

Certainly, they tend to show that the common view attributing much of the Prussian victory over France to the excellence of her railway organization is wrong. Given the shortcomings of their network and the errors in their doctrine, the Prussians must have owed their triumph either to some exceptional gift for improvisation that enabled them to overcome both problems during the campaign, or to the fact that railways did not, after all, have much influence on their conduct of the war. Which of these two interpretations is correct, it will be the task of the following pages to investigate.

Railways against France

On 13 July 1870, Bismarck released the edited version of the Ems telegram and two days later Prussia, together with the remaining German States, found itself in a war against France. Having expected something of the kind to happen, the Prussian army was ready; at the outbreak of war, it was only necessary to push a button in order to set the whole gigantic machine in motion.

Between August 1866, when the war against Austria came to an end, and July 1870, Moltke frequently addressed himself to the question of mobilizing and deploying his troops against France. Since the French army – unlike the Prussian one – was a standing force, it could be expected to be quicker off the mark. Assuming that Napoleon would exploit this advantage by launching an early offensive, the problem facing Moltke was not so much to prepare for an advance into enemy territory, as to employ the railways in such a way as to achieve numerical superiority at the earliest possible moment. The further forward the deployment was carried out, the less German territory it would be necessary to give up. Nevertheless, there were compelling reasons for concentrating the army well to the rear. In the first place, it was necessary to reckon with the possibility that a French offensive would disrupt the German railway network near the frontier. More important still, the number of trains that could be run over each line per day stood in an inverse ratio to the length of the section to be covered. Hence, the desire to hold as much German territory as possible clashed with the need to concentrate the army at the greatest speed, and it is typical of Moltke that his final plan (apparently prepared in the winter of 1869–70) provided for the concentration to take place far to the east behind the river Rhine, thus sacrificing, if necessary, the entire Rhineland to an almost unopposed French advance.[36]

As for the actual distribution of the troops, it was dictated less by strategic considerations and by what was known of the enemy's intentions than by the physical configuration of the railway system. As in 1866, the demand for speed made it imperative to exploit the greatest possible number of lines; hence the decision to deploy the 13 corps into which the Prussian army was now divided on a very broad front all along the Franco–German border. In this way, six lines were made available for the Prussian forces and three more for their south German allies, and not more than two corps had to share a single line.[37]

In the actual transportation, priority was given to the combat troops, those of the two corps on each line following one another directly, while their transport and services were supposed to come up later. This arrangement was logical, but it meant that the logistic instrument, the fragility of which had already been demonstrated in the previous campaign, was thrown out of gear before the war against France even started. More significant still, the burdening of the railways by trains carrying troops made it impossible to push supplies forward, and when trains carrying subsistence finally started running on 3 August the lines quickly became blocked.[38] The result was that, as in 1866, the supply services could not even begin to tackle the task of feeding the troops in their areas of concentration. The General Staff had ordered field ovens to be set up at Cologne, Koblenz, Bingen, Mainz and Saarlouis, fed from peacetime magazines, and supplies were purchased in Holland and Belgium and shipped down the Rhine. But as the troops had been separated from their transport, these supplies could not be distributed, and when complaints were made to Moltke he replied that quartermasters 'should limit themselves to what is strictly necessary and avoid bothering the railway authorities'.[39] Under these circumstances, the Army commanders were forced to use their own initiatives. Requisitioning transport at 400 wagons per corps, they purchased victuals on a grand scale because they were afraid, rightly as it turned out, that the experiences of 1866 would be repeated and the General Staff prove unable to cater to their men's needs.[40] Since the troops were at this time being fed by quartering, it was not surprising that shortages soon developed even though the country was rich and the population ready for sacrifice.

When the German advance into France started on 5 August, their I Army had as its line of communication railway line F; II Army had lines A, C, B and D (the latter in common with III Army) and III Army relied on lines D and E. To a large extent, however, these arrangements had already ceased to function, for the stream of supplies from the rear was such that the railheads, especially those behind II Army in the Palatinate, were becoming blocked even before the deployment was completed.[41] Once the advance started, moreover, the railways were quickly left behind and contact with them was lost. For example, it was not until after the battle of Spicheren on 6 August that the railhead behind I Army was advanced as far as Saarlouis, still inside German territory. The railhead of II Army was advanced to Saargemund on 11 August, then, four days later, to Pont-à-Mousson where it was destined to remain until the mobile phase of the campaign was all but over. III Army's railhead was at Mannheim, and this too could not be advanced beyond Mars-la-Tour.[42] Thus, the experiences of 1866, when the railways had been left too far behind to exercise any influence on the campaign, were repeated, with the one difference that, profiting from the lessons of the previous war, the Prussians had made preparations to construct an emergency railway round the fortress of Metz and were able to achieve something of a *coup* by the speed with which these preparations were translated into practice.[43]

While the fact that congestion similar to that of 1866 was allowed to recur points to the conclusion that German railway organization still left something to be desired,[44] the difficulties experienced in the rapid advancing of the railheads were not due to any lack of foresight. In countless memoranda covering the years 1857–70, Moltke had concerned himself with the possibilities of demolishing and rebuilding railways in wartime. In 1859, he had ordered the first *Eisenbahntruppe* formations to be set up, whose task it was to deal with these novel aspects of the military art. The idea behind their creation was that they should carry out minor repairs and build small bridges, work of a larger scope being left to civilian experts. The railway troops formed part of the pioneer force and were expected to guard the lines as well as restore them. Although the former task was formally taken away from them in 1862, it was found that, in practice, commanders often refused to allocate other units to this purpose.

In the war of 1866, there were three *Eisenbahnabteilungen* –

each consisting of a commander and 50–100 men, including 10–20 specialists – who were attached to the three Armies operating in the field. All served tolerably well, but proved wholly inadequate for the magnitude of the task. In particular, labour was lacking to complete the line Berlin–Görlitz, as well as to build emergency lines round the fortresses of Josephstadt and Königgrätz. All supplies had to be run over the single line Dresden–Görlitz–Reichenberg–Turnau, with results that have been described in the previous section.[45] In 1870, the number of railway-detachments had been increased to five (including a Bavarian one), each just over 200 men strong and commanded by a *höherer Eisenbahntechniker*, who also served as adviser on railway questions to the *General-Etappeninspekteurs*.[46] In the early stages of the campaign, however, these preparations proved not so much inadequate as irrelevant. Well-trained and well-equipped as the railway troops might be, they were powerless to do anything against the French fortresses barring the way, particularly that of Toul. True, as one historian has written, reducing this and other fortresses 'was only a matter of time and concentration',[47] but in spite, or perhaps because, of this it was not achieved until 25 September, when the French regular armies had already virtually ceased to exist and Moltke's forces were approaching the gates of Paris. That the reduction of Toul was not allowed higher priority was, therefore, itself an indication of the fact that the Prussians found it possible to make do perfectly well without any great need for the railways.

Meanwhile, after a somewhat muddled start, the Franco–Prussian War was developing into one of the most spectacular campaigns of all time. After Spicheren, there came Froeschwiller and Vionville–Mars-la-Tour, following which Bazaine with 160,000 French troops found himself penned up at Metz. Having won this victory, the German II Army was divided in two; four of its corps were left behind to invest Metz, while the remaining three were designated the Army of the Meuse and sent to the northwest in order to help fight the other part of the French army, now concentrating around Sedan. There followed, on 18 August, the battles of Gravelotte and Saint-Privat. By 1 September, Napoleon III had been surrounded at Sedan, and two days later he surrendered. The way to Paris was now open.

Having been compelled by the configuration of the railway network to open the campaign on a broad front without any very

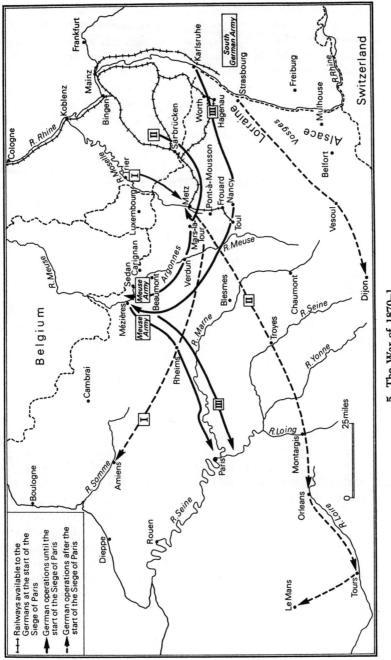

5. The War of 1870–1

clearly defined *Schwerpunkt*, the German army had now become grouped into two great parts. One of these, consisting of I and II Armies, was besieging Metz, where, although Moltke's foresight made it possible to advance their railhead to Remilly on 23 September, very great difficulties were experienced in keeping this stationary force supplied. Further to the north, III Army and the Army of the Meuse were preparing to follow up their victory at Sedan in order to march on Paris, despite the fact that, as the railways had been opened only as far as Nancy, they had almost totally lost touch with their bases of supply. The two halves of the German army being separated by the Argonnes, communication between them was very difficult, and giving mutual support even more so. That Moltke could still, under these conditions, order a further penetration hundreds of miles into enemy territory shows how little dependent on regular supply from base the German armies still were.

As King Wilhelm's troops started marching on Paris, chaos on the railways behind them was swiftly assuming monumental proportions. This was due to many different factors, including the continued attempts of supply agencies in the rear to rush supplies forward without regard to the ability of the unloading stations to receive them; a shortage of labour for the quick emptying of trains; the inability of the army's horsedrawn transport columns to clear the stations of the goods that kept piling up; and the tendency of local commanders to impress railway wagons as convenient temporary magazines. As a result, lines became blocked for hundreds of miles, the backlog of motionless trains stretching back as far as Frankfurt and Cologne. Though frantic efforts were made to clear the mess by unloading all goods regardless of the facilities available to store them (which, of course, meant that much of them was simply left to rot), on 5 September there were standing on five different lines no fewer than 2,322 loaded wagons, containing 16,830 tons of supplies for II Army alone. The traffic jams also contributed to an acute shortage of rolling stock, which led Moltke to write to the Army commanders on 11 September and ask them to use their cavalry in order to lay hands on French wagons and locomotives.[48] Soon after the start of the campaign, difficulties were being caused by *francs-tireurs* who attacked trains and obstructed the lines, and on 12 October Moltke wrote that it would take months to repair them at the places where they had been demolished.[49] Thus, although traffic was by no means very heavy – an average of only six supply trains per day for the whole army went to France during the first months of the war – congestion was such that, from 1 to 26 October, only 173 out of 202 trains sent from Weissenburg to Nancy ever reached their destination.[50]

The chaotic situations on the railways in the rear, however, did not prevent the Prussian armies from continuing their inexorable advance into France. On 6 September, the march on Paris got under way; by the end of the month, the ring around the city was in the process of being closed. Further in the south, Metz delayed the Germans until 24 October, when Bazaine's surrender made it possible for I Army to march west in order to join the siege of Paris while II Army was sent further south into the Loire Valley. During this entire period the railheads remained, as they had been ever since the end of August, around Nancy.

In the event, it was not until December 1870 – when Moltke's forces had extended south to Dijon, southwest to Orleans, and west to the English Channel – that the situation on the railways in their rear started improving. Three lines were now running across the frontier from Germany into France, but one of them – passing from Mulhouse through Vesoul and Chaumont to Paris – remained blocked by the French fortress of Belfort until the end of the war. Another line reached Paris through Metz, Mézières and Rheims, but this was blocked by no less than three fortresses and did not become available to the Germans until the fall of Mézières on 2 January 1871. These facts limited all German rail transport to a single stretch of railway between the Moselle Valley at Frouard and the Marne Valley at Blesmes, though even this line did not become available until the fall of Toul on 25 September, by which time the German armies had already reached Paris, and then demolitions of bridges and tunnels in the Marne Valley delayed the advancing of the railhead by another two months. The state of the line of communications behind II Army in the Loire Valley was even more difficult. Until 9 December the railhead remained at Chaumont, for demolitions in the valleys of the Seine and Yonne had rendered the lines further west impassable. Later, the railhead was advanced to Troyes, but by this time II Army had got still further away to the Côte d'Or. This particular railway was also exposed to numerous attacks by *francs-tireurs*, with the result that, towards the end of November, an entire army corps had to be detailed for its guard.[51]

Though 2,200 miles of French track were being operated by
the Germans when the war came to an end, traffic on them
always remained chaotic and sometimes hazardous. Trains were
involved in crashes, were derailed, and fell into the Meuse. Sometimes this was due to sabotage, but in most cases the accidents
stemmed from incomplete and hurried repairs, the inexperience
of the German personnel and slack discipline. Nevertheless, the
Germans tended to attribute every failure to the action of sabo-
teurs, which led to Moltke's notorious order that French hostages
should be taken along on the locomotives.[52]

There is no doubt that the German siege and bombardment of
Paris, involving as they did the concentration in a small space of
very large masses of men and heavy expenditure of artillery ammunition, would have been wholly impossible without the railways. Also, the view that the German use of the railways to deploy their forces at the opening of the campaign as a supreme
masterpiece of the military art is amply justified, though we have
seen that this triumph was only achieved at the cost of disrupting
the train apparatus before the war against France even got under
way. Between these two phases of the struggle, however, the
railways do not seem to have played a very important role, partly
because of difficulties with the lines themselves and partly
because of the impossibility of keeping the railheads within a reasonable distance of the advancing troops. Most surprising, however, is the fact that none of this had much influence on the
course of operations, or indeed caused Moltke any great concern,
which can only be understood by examining the actual methods
by which the German army of 1870–1 was fed.

Logistics of the armed horde

Ever since the Franco–Prussian War, historians have regarded the
supply organization as one of the Prussians' greatest achievements,[53] a belief that Moltke himself helped create when he wrote
that, in the entire history of warfare, no army had been as well
fed.[54] It is true that the German forces did not suffer from grave
supply difficulties during most of their campaign against France,
but it is not true that this fact was due to any superb feats of
organization. Failure to recognize this has tended to distort, not
merely accounts of the war itself, but also most attempts to allocate it a place as a stage in the development of the military art.

The shortcomings in the supply apparatus that were revealed by the 1870 campaign were not, admittedly, due to any lack of organization or foresight. From a very modest beginning, the train service of the Prussian army had by this time developed into an impressive machine, each corps being served by a train battalion with 40 officers, 84 doctors, 1,540 men, 3,074 horses and 670 wagons. The marching order of these was not rigidly fixed, but the combat troops could expect to be followed closely by the battalion's spare horses, pack horses, medicine cart and mobile canteen, all of which were supposed to stay with the battalion even during days of combat, and were known as *Gefechtsbagage*. Next came the so called small baggage, consisting of the wagons of the divisional staff, those carrying infantry ammunition, the field forges, the remaining canteen wagons, the troops' provision columns, plus one reserve provision column and one field hospital per infantry division. Finally, there was the heavy baggage; this consisted of ammunition columns, officers' baggage, the field bakery, the field hospitals not forming part of the divisions, the remaining provisions-columns, the pontoon-column, the second echelon ammunition columns and the remount depôt. Replenishment of the troops' vehicles was to be carried out when half of them were empty, for which purpose they were to be left behind – to enable the train columns proper to catch up with them – or driven to the rear during the night, in order to avoid crossing other units on their way to the front.[55] To help the corps in emergency, each Army also acquired a reserve wagon-park of several thousand vehicles by requisitioning. On paper, these were impressive arrangements. From the beginning, however, they failed to function, and for this Moltke himself was largely to blame.

As we saw, Moltke's swift deployment of his forces on the Rhine was only achieved by separating the troops from their transport, with the result that, when the campaign got under way, the latter had not yet arrived in the areas of concentration and was unable to discharge its functions. Given the relative speed of movement of troops on foot and transport by horse, especially since the latter was supposed to shuttle between the front and the railheads or at least stay in place in order to be replenished, this gap was not easy to close. Therefore during their advance to the frontier, the German troops had to be supplied by quartering and purchase, a procedure that caused friction and hardship to the civilian population. When the German–French frontier was crossed, the trains had still not succeeded in catching up. For example, those of III Army only reached the front in mid August, after several battles had already been fought and won, and German troops had crossed the Meuse.

During the advance into France, the supply problems faced by the various German forces were very dissimilar. On the left, I and II Armies marching against Metz did not have very far to go and could be kept within reasonable distances from the railheads. These, however, were too congested to be of much use, with the result that, even though three trains per day were supposed to arrive for II Army,[56] both Armies had in fact to be fed mainly by requisitioning, supplemented by captured French supplies.[57] This worked well enough as long as the German troops kept on the move. Once operations came to a halt around Metz, however, 'enormous difficulties' were experienced. The distance from the railheads now amounted to some forty miles, part of which were crossed only by narrow mountain roads. Congestion on these was extreme, and when they were weakened by rain a shortage of labour made it impossible to have them repaired. Having finally settled down, the troops around Metz were required to give up part of their transport to help the forces besieging Paris, which, at least, went some way to solve their supply problem because, as always, fodder had proved especially difficult to procure and the number of dead horses was legion.

Though it will be remembered that careful preparations had allowed a railway to Remilly to be built quickly, this was nullified by the fact that congestion forced supplies to be unloaded wherever labour and space were available, which was often in stations far to the rear. Provisions were unloaded without regard to storage facilities and left to rot, and since no labour was available to bury them (the local population proved reluctant to carry out this task, nor did it possess the necessary tools) the stench soon rose high. These difficulties were just beginning to be overcome when, following the battle of Sedan, I Army and the Army of the Meuse started obeying Moltke's order to send back captured rolling-stock, with the result that the station of Remilly was congested still further.[58] All in all, keeping I and II Armies supplied during the siege of Metz therefore proved considerably more difficult than it had done during the previous and, as we shall see, the subsequent period.

Meanwhile, on the left flank of the German advance, III Army was free from these difficulties because it lived entirely by requisitioning. Its supply trains had only just succeeded in catching up with the troops when an unforeseen strategic opportunity arose, as Macmahon's army was reported standing between Rheims and the Meuse on the Germans' right. Faced with the choice of either plunging into the Ardennes without properly organized supplies in order to cut off the enemy, or waiting until the logistic situation improved at the risk of allowing him to escape, III Army decided to march. A rest-day that had been ordered for 27 August was summarily cancelled, and the troops were instructed to look after themselves by living off the country, supplementing their finds with iron rations when necessary. This was, of course, the correct decision. Shortages of victuals did appear towards the end of the month, but this was a small price to pay for the victory at Sedan.[59]

The experience of the other force which took part in the battle of Sedan, the Army of the Meuse, was somewhat similar. When it had been decided to send four corps of II Army to the northwest, a reserve to supply them for fourteen days was built up at Mars-la-Tour. However, the Army was unable to bring this up because too many wagons had been given up to transport troops, or to help with the construction of the railways. Consequently, the Army during its march had to be fed by the usual combination of quartering and free purchase, and while these sufficed on the whole, shortages did arise during the last days of August when it became necessary to resort to the iron rations.[60] Once again the victory justified the decision, though both German Armies were marching without any reserves and would have faced disaster if the battle had gone against them.

With the heavy concentration of troops around Sedan during the last days of August, supply difficulties naturally arose, though these were alleviated by the fortunate capture of French stocks at Carignan. Not until after the battle had ended did the supply columns of the two German armies succeed in catching up with them, though by this time it was found that their load had been much reduced through their own consumption. It was impossible, therefore, to build up a proper basis for the advance on Paris, the more so since the nearest railhead was still at Metz, some 80 miles to the rear. These difficulties were made good, however, by the fact that a French army capable of opposing the German troops no longer existed. Consequently, the latter were able to spread out over a broad front, with each corps marching over a separate

road. The country being very rich, no great difficulties were experienced, and it even proved possible to form a surplus at
Rheims and Châlons. However, the closer the Germans came to
Paris, the more often villages were found deserted, crops burnt,
and cattle driven off.[61] Similarly, I Army after the fall of Metz
set out on its way towards Paris through the valley of the Seine
without making any special logistic preparations. Reliance was
again placed on requisitioning, though in this case results proved
disappointing and it became necessary to resort to free purchase
on a grand scale.[62]

Alone among the four German Armies, II Army did make some
considerable preparations prior to its march from Metz to Orleans. Although the siege of Metz lasted for about two months, the
difficulties of feeding the investing troops were finally overcome.
In mid-October, ample provisions were available around the city
despite the need to feed, in addition to the German forces, 150,000
French prisoners. This enabled the *Intendantur* of II Army to
issue an order that all units were to start their march to the
Loing with their provisions-columns fully loaded, the intention
being to replenish them from the rear in order to arrive with these
stocks still intact. To this purpose, the Army was able to take
along a total of 4,750 tons of food and fodder; at seven pounds
per man per day, this should have sufficed to supply 100,000 men
for 17 days.[63]

In the event, only the first part of these plans proved capable
of realization. Around 20 October, II Army departed from Metz
with its provisions-columns fully loaded. The transport convoys
earmarked to replenish them from the rear, however, were hopelessly unable to keep up with the pace of the advance, so that
requisitioning had to be resorted to on this occasion also. As the
country in question had already been scourged by the French in
order to victual Paris, it was decided that this could be best
achieved by paying in cash, which was acquired by laying the
towns on the way under contribution in exactly the same manner
as Wallenstein had done two and a half centuries earlier. Thanks
to this lack of respect for private property, the Germans did in
fact reach the Loing with their provisions-columns full. Eventually, even the formality of paying for the requisitioned goods was
abandoned as the French built their 'national' armies and the
war assumed a less chivalrous, harsher character.

Having crossed the Loing with its provisions-columns full or replenished, II Army continued its march to the Loire through extremely rich territory, though some difficulties were caused by the fact that, since many of the inhabitants had been called up to serve in Gambetta's new armies, labour to bring in the harvest was short. However, the potato crops had already been gathered and they were abundant. Meat and vegetables were also plentiful, and when any problems arose it was always possible to supplement the men's rations with the famous *Erbstwurst*. Throughout this period, II Army was more or less out of touch with the railways. Hopes for the rapid transfer of the railhead from Blesmes to Montargis had been disappointed, and even when it was finally advanced to Lagny at the end of November the distance to be covered by road still amounted to some 130 miles in both directions, so that the troops preferred to help themselves.

While II Army was thus engaged in a leisurely tour through the heart of France, very great difficulties were experienced in supplying the three German Armies now concentrating in a small area around Paris. Under conditions of static warfare, requisitioning soon ceased to be satisfactory, and so far away were the railheads that the wagons of III Army, for example, took ten days for the return journey. Moreover, all the Armies had lost much of their transport – that of I Army was down to exactly *one per cent* of its original establishment – so that not enough vehicles were available to bring up even half of the quantities consumed each day. As usual, difficulties were experienced with the railways, which in this period were constantly being blown up in addition to being burdened by the transportation of heavy artillery for the bombardment of Paris, which Moltke had ordered on 9 September.

To solve the problem of subsistence, which was greater by far than anything experienced since the beginning of the war, the German forces around Paris were turned into a gigantic food-producing machine, the like of which had not been seen on the battlefields of Europe since the end of the eighteenth century. Thousands of soldiers were taken away from their posts in order to gather in the harvest (corn, potatoes, vegetables) and to process it by means of local machinery, such as threshing machines, mills, bakeries etc. Regular markets were established and kept supplied by the French peasantry. To supply the troops with water, a river was diverted. The army had therefore been made largely self-supporting, and for this reason could not find the time or the resources to engage on its proper business, war. Not

Supplying War

until the end of November were the railheads pushed forward sufficiently to relieve the wagon-park and make it possible to embark on the stockpiling of ammunition for the bombardment of Paris.

If, during the greater part of the war against France, the German armies lived off the country and did not depend heavily on a supply of food from the rear, consumption of ammunition was so small that an instrument to care for its replenishment was hardly necessary at all. As in 1866, expenditure of infantry ammunition was so low that the reserves carried with the troops were only partly consumed. Thus, in five months' campaigning, an average of only fifty-six cartridges were fired by each Prussian *Landser*, which was rather less than what he carried on his back and only about one third of the stocks available in the corps' organic transport. If temporary shortages nevertheless occurred (above all at I Army during the battle of Mars-la-Tour) this was due, not to any shortage of ammunition, but to the inability of the troops' wagons to carry it forward during the battle.[64]

The figures for artillery ammunition are as follows:[65]

Date	Battle of:	No. of guns:	Rounds expended:	Average per gun:
4 August	Weissenburg	90	1,497	16
6 „	Wörth	234	9,399	40
14 „	Borny	156	2,855	18
16 „	Mars-la-Tour	222	19,575	88
18 „	Gravelotte	645	34,680	53
30 „	Beaumont	270	6,389	23
31 „	Noiseville	172	4,353	25
1 Sept.	Sedan	606	33,134	54

Since not all the guns took part in all battles, however, overall expenditure was much lower than these figures indicate, amounting to an average of only 199 rounds per gun for the entire war.[66] The normal number of rounds carried inside each corps was 157, therefore the turnover was so low that the troops could afford to disregard the entire *Etappen* system and refill their vehicles directly from the railways, even though the distances involved were sometimes very considerable. Moreover, so small was consumption that the troops did not hesitate to part with their vehicles for weeks on end. For example, after Sedan, the wagons

of III Army were sent back to Nancy to refill, and did not rejoin
the corps until after the latter had reached Paris.[67] It is clear,
therefore, that in this respect also Moltke's army did not depend
on supply from the rear to any very great extent. Rather, as in
the campaigns of earlier days, a very large proportion of the ammunition required was simply carried along at the outset, with
the result that the army was self-contained for the greater part of
the war.

Did wheels roll for victory?

In the annals of warfare, the operations of 1870–1 are often said
to occupy a special place, because they were followed by 'a
modern line of communications, stretching away from the
[troops] to their base' and served by a 'meticulously organized'
train apparatus. This apparatus is said to have been 'so inti-
mately interlocked with the force it served that any separation of
the formation from its own particular sources of supply. . .must
spell dislocation in its movement and may mean disaster'.[68] In
addition, it was the first time – in Europe, at any rate – that the
full potentialities of the railways as an instrument of war were
realised, the beginning of a process, in other words, which gradually took away 'the secret of strategy' from the soldiers' legs and
transferred it to wheels instead.

As is shown in the above pages, these claims are entirely without foundation. That they have nevertheless been accepted for so
long, is a remarkable testimony to Moltke's ability to impress his
own account of events upon history, and, to an even greater
extent, to the credulity of historians and their readiness to accept
without question the words of a commander whom fate has
crowned with victory, in spite of the fact that all the evidence to
the contrary has long since been published and is readily available.

A detailed analysis of the shortcomings of the Prussian army's
logistic system will take us too far and can, in any case, merely re-
peat what has already been said in the preceding pages. Nevertheless, the following points appear worth making:

(1) While the Prussian army did, in 1870, have a supply service theoretically capable of catering to its needs, this service
proved an utter failure in practice. Though marching performances were not terribly high – the pace of the advance seldom

averaged more than ten miles a day for a fortnight at a time – the method of deployment had thrown the train apparatus out of gear even before the campaign started. Train troops were insufficiently armed, and thus unable to defend themselves. Marching discipline was slack, and repair facilities for vehicles so inadequate that nine out of every ten wagons had to be left behind.[69] As a result, it soon became clear that the hopes pinned on the train were incapable of fulfilment. Not merely the provisions-columns, but the entire elaborate organization of mobile field bakeries and butcheries failed to work, making it necessary for the troops to help themselves on the great majority of occasions. Indeed, so irrelevant were the trains to the army's supplies that field commanders were indifferent to their whereabouts, with the result that they frequently remained without orders for weeks on end and finally had to go looking for their units on their own initiative.[70]

(2) As regards the supply of ammunition also, the success of the campaign was due less to any elaborate system of replenishment from the rear than to the fact that expenditure was very small throughout the campaign. This more than made up for the fact that, at the outset, estimates concerning the relative expenditure of infantry- and artillery ammunition proved to be wrong.[71] On the whole, despite shortages here and there, providing the troops with ammunition was a very much easier task than keeping them fed. In this sense, as in so many others, the campaign of 1870 cannot be regarded as 'modern'.

(3) The role of the railways during the war has been grossly over-estimated, most historians being all too ready to accept the extravagant claims made by Moltke and his 'demigods' as to their importance.[72] In fact, the railways fulfilled a crucial function only during the period of deployment, following which they ceased to play a major role until well after the mobile phase of the campaign was over and the war all but won. This was due, partly to difficulties with the railway traffic itself, partly to the inability of the railheads to keep up with the advance, and partly to the impossibility of moving supplies from them to the front. Altogether, these three factors made it possible for the railways to play a role only when operations were more or less stationary, and even then – as around Metz – the greatest difficulties were experienced in their operation.

An interesting aspect of the railway problem is the failure of

Moltke and the General Staff to learn from experience. Every one of the obstacles that arose in 1870 had already been rehearsed in 1866, and yet they were allowed not only to recur but to become infinitely worse. In part, this was due to the inherent limitations of a system of supply based on the unfortunate combination of the technical means of one age – the railways – with those of an earlier one. It is no accident that the worst difficulties occurred at the transfer points from one system to the other, i.e. at the unloading stations. However, errors in organization did also contribute to the confusion, and for this Moltke was undoubtedly to blame. Enough troops to guard the lines against sabotage were not available, and difficulties seem to have been experienced in the transfer of rolling-stock from the German civilian railways to the army.[73] Despite the experiences of 1866, a central supply- and railway-transportation headquarters for the whole army had not been created, with the result that the contractors, in their anxiety to make as much profit as possible, pushed forward the maximum quantity of supplies without regard to the limitations of the railways.[74] Labour and vehicles to help in the unloading of trains were short. German troops frequently dismantled the signal and communications gear of captured lines, an evil that successive orders were unable to eradicate. Consequently hundreds of thousands, indeed millions, of rations were allowed to rot away. This was in spite of the fact that the demands made on the railway network were, in reality, very modest. Consumption of each corps being only about 100 tons a day (fodder included), the entire requirement of the Prussian army could have been covered by six or seven trains. This was well below the capacity of one well-organized single-tracked line.

More significant than the above, however, was the fact that the railways were entirely unable to keep up with the pace of the advance. This, as we saw, was due less to any lack of preparation for their restoration than to the fact that the *Eisenbahntruppe* were helpless against the French fortresses blocking the lines. That Moltke had failed to foresee this is difficult to believe. He had, after all, written a memorandum on the influence of the Austrian fortresses on the events of 1866, and the rapid reconstruction of a railway around Metz was a military masterpiece. Rather, the failure to draw conclusions from the events of 1866 must have been due to the fact that, contrary to what is generally thought, the railways did not play a major role in the war against

Austria. Unlike modern historians, Moltke recognized this fact, and it led him to the logical conclusion that his order of priorities was right. This, of course, was proved to be the case in 1870. The advancing Prussian forces simply bypassed the fortresses, and not until the mobile phase of the campaign was over did the latter turn into serious obstacles.

Another aspect of the war of 1870 about which many writers have been mistaken is the alleged superiority of the German railway system over the French one. Exactly how this erroneous opinion originated I am unable to discover, as there is not the slightest evidence in its favour. Since the German railways are supposed to have played such a large role in the victory, most historians seem to assume that they *must* have been superior in some way. This, however, was not the case. On practically every account, the French railway system in 1870 was actually better than the German one. This was even more true of the military aspect than of the civilian one, for political factors made it easier for France than for Germany to take strategic considerations into account. During the 1860s, therefore, it was thought that nowhere did the French enjoy as great an advantage as on precisely this point.[75] In any case, as had been recognized long previously, the distinction between a 'civilian' and a 'strategic' railway network is largely unreal:[76]

> While every State will undoubtedly hesitate to undertake the expense [of constructing military railways]. . .it will naturally try to accelerate the traffic of people and goods in. . .the directions dictated by the density of the population and the volume of trade. However, these directions, these routes of traffic and commerce, are usually – indeed, almost invariably – identical to the lines of operations of armies.

That these sensible words could be written by the Bavarian minister of war in 1836, when experiments with the military use of railways had hardly begun, is a sad testimonial to the readiness of many historians to copy each other's words without giving the slightest thought to the evidence on which it is based.

It is true that great difficulties were experienced by the Germans in using the railways for military purposes during the campaign, yet the problems experienced in transporting supplies from

the railheads to the troops were probably greater still. To overcome them, some attempts were made to use road steam-
locomotives, but without much success. Since the engines were
only able to use good roads, they were frequently forced to make
large detours. Their ability to negotiate slopes was limited, and
they were slow and difficult to manoeuvre. Road locomotives
were used to bypass the fortress of Toul, to circumvent the
demolished tunnel at Nanteuil on the railway from Nancy to
Paris, and to go round the destroyed bridge at Donchery (this
attempt broke down completely, the cargo finally reaching its
destination with the aid of forty-six horses, eleven drivers and
twenty-five pioneers). There were seven other attempts, all
marked by breakdowns and a tremendous waste of time, since the
loads transported, usually railway locomotives, were too heavy to
be carried in one piece and had to be dismantled, transported and
finally reassembled at the point of destination. That these experiments did not give rise to any great enthusiasm is understand-
able.[77]

Given the failure of the train service to discharge its proper
function, the entire German campaign of 1870–1 was only made
possible by the fact that France is, after all, one of the richest
agricultural countries in Europe, and that the war started in a
favourable season of the year. For much of the mobile phase of
operations, the Germans were therefore able – indeed compelled,
by the failure of their supply apparatus – to live off the country
much as Napoleon's soldiers had done seventy years earlier. In
this, they were helped by the fact that Europe had grown far
richer since 1800, and if eighty people per square mile were considered very good in 1820 the average in 1870 was nearer 120.
The size of armies had of course also risen, but this was largely
offset by the fact that all of Moltke's forces in 1870 were never
concentrated at any single point and were able, following the
French defeats early in the campaign, to spread out their corps
over a wide front. Thus, during the march of II Army from Metz
to the Loing, it was estimated that the country traversed had no
less than 100,000 tons of flour and 100,000 of fodder in store,
against a consumption by the army during its passage of only
1,080 and 5,500 tons respectively. As the intendant of II Army
wrote, 'in the enemy's country it is unnecessary to economize as
much as at home'.[78]

Though it was, therefore, usually possible to feed the Armies from the country, this was only true as long as they kept moving. When operations came to a halt, as they did during the sieges of Metz and Paris, very great supply difficulties were at once experienced. In the case of Metz, these were only solved after a considerable period of time by the great efforts of the train companies, aided as they were by the fact that the railheads were relatively close at hand. In the case of Paris, it was necessary virtually to suspend the military functions of the army for the duration of two months and have the troops look after their provisions instead. This employment of an army as a food-producing machine is, to my knowledge, unique in the annals of war after 1789 and would have caused much amazement to Napoleon. Certainly, it was the last occasion when a large force belonging to an advanced State was so utilized. It is the fact that Moltke's forces could only live as long as they kept moving, and experienced the greatest difficulty in staying in one place for more than a few days at a time, that the supreme proof lies that the military instrument in his hands did not, after all, belong to the modern age.

4

The wheel that broke

State of the art

For Europe as a whole, the period between 1871 and 1914 was one of very rapid demographic and economic expansion. In just forty-four years, population grew by almost seventy per cent, from 293 to 490 millions. During the same period, industry, trade and transportation developed by leaps and bounds until, on the eve of World War I, they had totally transformed the face of the continent. In 1870, the combined production of coal and lignite by the three leading industrial countries – Britain, France, and Germany – amounted to just under 160 million tons a year; by 1913, it had more than trebled to reach 612 million tons. Similarly, in 1870, the production of pig iron in the same three countries was around 7.5 million tons a year, whereas by 1913 it had grown to 29 million tons, an increase of almost 300 per cent. This expansion, needless to say, was accompanied by vast changes in the pattern of occupation and residence. If the industrial revolution may be said to have begun a hundred years before 1870, it was nevertheless the Franco–Prussian War that truly ushered in the age of coal and steel.

As factory chimneys grew ever taller, so did the size of the military instruments maintained by the major continental powers. In fact, the expansion of European armies and navies during the period under review, particularly in its second half, was even more rapid than that of population and industry. Social progress, increasing administrative efficiency, and, above all, the now almost universally adopted principle of conscription, made it possible to raise huge forces which, in relation to the size of the politico-economic systems supporting them, were far larger than anything previously recorded in history. For example, in France, the second-largest military power, the pool of trained military manpower available in 1870 amounted to not quite 500,000 men in a population of thirty-seven millions (a ratio of about 74 to 1). In 1914, however, this total had grown to more than four millions, despite an increase in population of less than 10 per cent. Similarly, though the population of the German Empire grew by almost two thirds in this period, the expansion of the armed forces was such that one out of thirteen people was immediately available for military service at the outbreak of World War I, as opposed to only one out of thirty-four in 1870–1. In Europe as a whole, the size of the armed forces in their various degrees of readiness and mobilization stood at about twenty million in 1914, a figure probably never again to be approached in time of peace.

As warfare became more complex the *impedimenta* carried by armies into the field, as well as their consumption per man per day, increased at an even greater rate than their manpower. To mention only a very few figures for the country that concerns us most, the wagons constituting the train (field bakeries, hospitals, engineering equipment, etc.) of a German army corps numbered thirty in 1870, but this had been more than doubled forty years later. The count of artillery pieces available to the North German Confederation for its war against Napoleon III is said to have stood at 1,584 whereas in 1914 the total must have been nearer 8,000, many of which were far bigger and heavier. Though the number of weapons of all types organic to each corps changed surprisingly little (the number of guns, for example, grew only from sixty-four to eighty-eight) those of 1914 were mostly quick-firing and sometimes automatic, capable of shooting off quantities of ammunition much greater than their 1870 predecessors. At that time, 200 rounds per rifle were carried along inside the various transport echelons (the body of the soldiers, battalion and regimental wagons, corps reserves) of each corps, but only fifty-six of these were, on average, expended during six months of campaigning. In 1914, the number of rounds carried had increased to 280, and these were completely expended during the very first weeks of war. In 1870–1, every German gun had fired an average of just 199 shells, but the 1,000-odd rounds per barrel held in stock by the Prussian War Ministry in 1914 were almost depleted within a month and a half from the initiation of hostilities.[1]

With the increased consumption of ammunition came the problem, largely novel in 1914, of replacing the weapons themselves. In 1870–1, as indeed in all previous periods, a gun was expected to last for the duration of the campaign and usually did, artillery-fire being seldom powerful enough to thoroughly destroy it. A carriage might be blown to pieces, but the barrels themselves were almost indestructible. By 1914 this situation had completely changed, as artillery fire was now easily capable of quickly reducing whole batteries to mangled heaps of twisted steel. As for the guns, so for all other pieces of arms and equipment, the regular replacement of which was to form a heavy and growing burden on the transportation services.

To meet these and other demands, the number of horses serving with armies in the field was constantly being raised, the proportion of animals to men increasing from about one to four in the Prussian army of 1870, to one to three for the same army forty years later. Horses, however, eat about ten times as much as men, the result being that, even though the quantities consumed by the troops themselves did not presumably undergo significant change, the total subsistence requirements per day, for any given unit, increased by about fifty per cent.

Man for man, the weight carried along, and consumed from day to day, by armies in 1914 was therefore many times that of 1870, over and above the increase in sheer size. To compensate for this extraordinary growth, the most important means of strategical transportation – recognized, since the early 1860s, as the railways – also underwent spectacular development. However, the limitations of railways did not escape the notice of military experts. Railways are by nature a rather inflexible instrument, though by 1914 the density of the European network was such that Moltke's dictum about an error in deployment being impossible to correct for the duration of the campaign was beginning to lose some of its force.[2] Though 117 trains could carry a 1914 corps over 600 miles of double track in just nine days, loading – and unloading – times were such that the use of railways for covering distances under 100 miles was regarded as uneconomical, at least for large units of all arms. The lines themselves, as well as the troops using them, were singularly vulnerable to enemy action. For all these reasons, it was difficult to use railways operationally, and their employment was confined mainly (though not exclusively, as the battle of Tannenberg was to show) to transportation to, and behind, the front.

Within these limitations, how well did the development of railways keep up with the growth of armies? There were 65,000 miles of track in the Europe of 1870, as against 180,000 in 1914. This represented an increase of almost 200 per cent, with the leading countries, Germany and Russia, having an even larger increase. Qualitatively, progress was even greater. At the time of the Franco–Prussian War it was reckoned that a single line could carry eight trains a day, a double one twelve, whereas on the eve of World War I the figures were forty and sixty respectively. In August 1870, nine double-tracked lines served to deploy 350,000 German troops in fifteen days, so that 2,580 men rode each line each day. Forty-four years later, thirteen lines brought up 1,500,000 men to Germany's western frontier in ten days, making 11,530 men per day per line. Also, wagons had become bigger and locomotives more powerful, so that it was possible to carry the subsistence of a corps for two days on a single train, which was half the number required in 1870, in spite of the fact that, in the meantime, the effectives of a corps had risen by fifty per cent from 31,000 to 46,000 men.[3] These figures are far from exhaustive, but they do tend to show that as far as the tasks of mobilization, deployment and supply were concerned, the development of railway transportation did keep up with the increase in the size and bulk of armies.

The same, however, was not true for transportation beyond the main-line railways. All armies had, it is true, developed light field railways in 1914 and trained units in their use, but the capacity of these was limited and they could hardly be regarded as anything but temporary substitutes.[4] Time and terrain frequently imposed strict limits on the construction of such lines, with the result that troop-movements, as well as the transportation of material and supplies, had to be carried out mainly by other means. Qualitatively, transportation had improved hardly at all; for their tactical mobility, the armies of 1914 were still dependent on those time-honoured means of locomotion, the legs of man and beast. In theory, there was no reason why marching columns should not be able to sustain a steady fifteen miles a day, a figure that had not changed since time immemorial. However, this was being made increasingly difficult by the huge proportions of the trains. Between 1870 and 1914, the number of wagons on the establishment of each corps had more than doubled, from 457 to 1168, and this was quite apart from the often still greater amount of transport required to refill the troops' mobile reserves as they became depleted. Though the horse-drawn transport operating in the army's rear was capable of considerably outmarching the men (twenty-five miles per day being reckoned as a steady average), it was forever shuttling forward and backward between front and base and was therefore certain to fall farther behind with every day's march. These factors, combined with the great increase in the quantity of supplies consumed, were responsible for the fact that the so-called critical distance, the maximum one at which a force could operate away from its railhead, was actually falling during the period under discussion. By the early twentieth century, the 100 miles of the 1860s had fallen to about half that number.[5] All such figures are dependent on a great many variables –the weather, the state of the roads, enemy interference with the transports, and the like – to be very meaningful; nevertheless, the downward trend is unmistakeable. To exacerbate the problem, in 1914 the combat troops of a corps took up so much road – twenty miles and more – that the transport companies often found it difficult to reach them in one day's march. A corps, in other words, was getting so big that it was difficult to keep it supplied even when it was not advancing at all! To this extent the mobility of armies had declined relative to their bulk during the years leading to World War I.

Logistics of the Schlieffen Plan

The debate as to why the German Army failed to conquer France in the campaign of 1914 started very early after the events, and has since been continued with almost unabated vehemence. Even before the year was out, some of the leading figures in the great drama had put on record their own versions of what had happened during that autumn. The controversy was carried on in a flood of memoirs published during the 1920s and 1930s, and when the participants finally withdrew, historians stepped in.[6] Such factors as the state of the road- and railroad-network in Belgium, the density of the troops per mile of front, and difficulties of marching distances and supply are mentioned in many of these accounts. However, on the whole, the logistic aspects of the Schlieffen Plan, both as conceived by its originator and as put into practice by his successor, have been thoroughly neglected.[7] Whether or not the Plan was logistically feasible; what part, if any, did logistic factors play in its failure; whether, finally, the hard facts of distance and supply would have enabled the German army to carry on had the battle of the Marne gone in its favour; all these are questions that remain to be answered.

As might be expected, the prime considerations governing the evolution of the Schlieffen Plan, from its origins in 1897 to its fully-developed version of December 1905, were not logistic but strategical. As the German chief of the General Staff saw it,[8] his country was surrounded by enemies on all sides and would sooner or later become involved in a war on at least two fronts. Working within the tradition established by Clausewitz,[9] Schlieffen aimed not merely at the more or less incomplete defeat of Germany's opponents but at their total annihilation. For a number of reasons – including the ratio of troops to space and the availability of a good road- and railroad network – he felt that this aim could be most easily achieved in France, which accordingly became the target for the bulk of the forces at his command. East Prussia was to be protected against the Russians only by a weak screen that was to hold out somehow until victory over France made it possible to bring up reinforcements. Thus, the whole Plan depended on speed in mobilization, deployment and execution, for which a total of forty-two days were allocated.

Acting against the possibility of a swift victory in the west was the fact that France had heavily fortified her border with Germany, which led Schlieffen to believe that the prospects of a successful breakthrough were distinctly unfavourable. An outflanking advance through Switzerland was considered and ruled out for topographical reasons.[10] This left a lunge through Belgium as the only possible alternative. As it finally crystallized, the Schlieffen Plan displayed breathtaking – not to say foolhardy – boldness. Some eighty-five per cent of the German Army were to be deployed on the Reich's western frontier, and of these 7/8 were to form part of the right wing, consisting of five Armies with 33½ corps (two more were to be brought up later from the left wing in Lorraine) and eight cavalry divisions between them. Echeloned from right to left, this mighty phalanx was to march west into Belgium, wheel south against France, envelop Paris from the west, and, leaving behind forces to invest the city, advance east and finally north-east in order to take the French Army in the rear and pin it against its own fortifications.

Schlieffen's great Plan has been faulted on both political and operational grounds,[11] but it is the logistic side that we are concerned with here. In this respect, the first question to be resolved was just how large the wheel through Belgium should be. Strategic considerations of speed and concentration demanded that the outflanking movement be made as short as possible, that is a thrust along the southern (right) bank of the river Meuse against the line Mézières–La Fère. However, it was feared that a manoeuvre on such a narrow front would find space too restricted, and roads too few, to carry the Army and allow it to deploy. Furthermore, on the size of the wheel depended the width of the area of concentration to be used by the right wing Armies, prior to the beginning of the campaign. Had the shift in the direction of the advance from west to south taken place near Namur, as seems to have been Schlieffen's original intention in the 1901–2 version of this Plan,[12] this area could not have extended northward further than Saint-Vit. Between that point and Metz – the pivot of the right wing – there were only six double-tracked railways coming in from the east – whereas the imperative demand for speed dictated, as it had in 1866, that the maximum number of railways be made use of. The net result was thus a conflict between strategic and logistic considerations, and Schlieffen resolved it in favour of the latter. To make the most extensive use of the double-tracked railroads leading to Germany's western frontier, he decided to detrain his troops all along the line from Metz to Wesel.[13] To enable the Army to advance without undue congestion he proposed to violate Dutch neutrality in addition to that of Belgium, by seizing the province of Limburg (the so-called Maastricht Appendix) as well as that of Northern Brabant.[14] Finally, in order to secure sufficient roads inside Belgium proper, he stretched the front of the advance until, in his own words, 'the last grenadier on the right wing should brush the Channel with his sleeve'.[15] This presented yet another advantage, for it would enable the Germans to 'scoop up' in their enveloping movement not merely the Belgian Army but any British force that might be coming to its aid.

If the size of Schlieffen's newly-prescribed wheel helped him solve one logistic problem it immediately created two others. First, since the time that could be allowed for the operation as a whole still stood at 42 days, tremendous marching performances had to be demanded of the troops. In particular, those forming the extreme right wing would have covered almost 400 miles by the time they reached the Seine way below Paris after passing

near the Channel coast. Second, the enormous size of the movement presented the problem of supporting and supplying large
forces – von Kluck's 1. Army on the extreme right alone numbered
well over a quarter of a million men – at such tremendous distances from the homeland. Schlieffen, in other words, resolved the
conflict between logistic and strategic considerations only at the
cost of creating two new, and, as it turned out, formidable sets of
logistic difficulties.

To 'solve' the first of these problems, Schlieffen simply wrote it
into his Plan that the troops of the right wing would have to make
'very great exertions'. Normally, a corps could be expected to sustain an advance of fifteen miles per day for three days on end, but
Schlieffen ignored this and instead provided for Kluck's Army to
cover the distance to the Seine in about twenty-five days, with
days of fighting not excluded. Such performances might, perhaps,
have been asked of Napoleon's *Grande Armée* in its prime, but
we have seen that the ponderous bulk of the armies of 1914 made
them singularly unsuited to such feats. The danger thus existed
that, when finally coming face to face with the French Army at
the end of their tremendous march, the German troops would be
too exhausted to give a good account of themselves, and in fact it
was exhaustion, as much as any other factor, that would have
prevented the continuation of the advance even if the battle of
the Marne had been won by Germany.

As for the second problem, Schlieffen does not appear to have
come to grips with it at all. Though his admirers have claimed that
'a warlord [sic] of Schlieffen's calibre would have carefully considered, prepared, and determined all the arrangements as
regards lines of communication, *Etappen*, railway traffic, and the
supply of ammunition, provisions and military equipment as a
matter of course',[16] there is in fact little to show that he was much
preoccupied with the question as to how the right wing was to be
sustained during its rapid and far-flung advance. An elaborate
three-tier supply system – based on the one first introduced by
the elder Moltke in the mid-nineteenth century – did exist. Under
this system, German infantry (but not cavalry, a point to which
we shall have occasion to refer later on), regiments and corps
each had their own organic transport columns, divided into two
echelons and marching either with the fighting troops or directly
behind them. These were replenished by the heavy transport companies operating in the zone of communications, which in turn

were fed from the railways.[17] But the system as a whole was rigid
and elaborate, better geared to a slow, methodical advance than
to a war of *sturmisch* movement. In particular, the wagons forming the system's second tier were liable to be left behind as soon
as the speed of the advance exceeded some twelve miles a day.
Even if, by some miracle, this did not happen, the range at which
they could support the Army was strictly limited. It must therefore have been clear to Schlieffen, as it was to the leading military
minds of the period, that in the long run it would not be possible
to move the Army any faster than the pace at which the railheads
could be made to follow in its wake.

There was normally supposed to be one double-tracked railway line behind each Army.[18] To seize, repair and operate them
quickly was a task of crucial importance, for which the *Oberste
Heeresleitung* (OHL) had under its own direct control some 90
companies of highly trained railway troops. Equipped with so-
called *Bauzuge*, trains carrying everything needed to repair
damaged tracks and lay new ones if necessary, these units were
expected to march with the leading forces or even precede them.[19]
Forecasting – correctly, as it turned out – that the French would
thoroughly demolish their railways in the general area of the
Meuse Valley between Verdun and Sedan, Schlieffen wrote that
'lines of communication must be sought mainly through Belgium
north of the Meuse'. How, one may well ask, did he expect to
supply the three Armies operating south of that river? And what
if the Belgians were to blow up the railroads in the northern part
of their country, the task being facilitated by the countless tunnels,
bridges and flyovers marking 'the world's best network'?[20] To
these questions, Schlieffen appears to have had no answer beyond
the dubious argument that 'the Belgian railways form the best
possible connection between the German and French systems'.[21]

The impression that the logistic side of the great Plan was not
properly thought out, gains in strength from the fact that
Schlieffen, who made the railway department of the General Staff
carry out extensive war-games to test the feasibility of transporting troops from one wing of the Army to the other, and from the
western to the eastern front, did not apparently use similar means
to examine the supply and maintenance of the troops on his all-
important right wing. The reasons for this indifference are hard
to discover. It may, as one writer thinks,[22] have stemmed from
the hope that the question would not arise in all its magnitude

during a short, victorious campaign such as the advance into France was expected to be. A more likely explanation, however, is that Schlieffen defied the common military wisdom of his time[23] in that he hoped to feed his Armies, in part at least, from the country they traversed.[24] As for the supply of ammunition and other equipment, it was still, in this innocent age, expected to be 'as nothing' compared to that of food and fodder.[25] To go by such evidence as can be found in the various drafts of Schlieffen's Plan, the logistic side of this intention appears to have rested on singularly shaky foundations. Exactly how large a proportion of their subsistence – and especially of that all-important commodity, fodder – the German Armies would be able to draw from the country depended on the season of the year and was therefore impossible to foresee. Nor were the tables laying down the consumption of ammunition, based as they were on experience forty years out of date and making completely unrealistic assumptions,[26] of much use to those responsible for planning the Army's transportation.[27] Just how little Schlieffen appreciated what the war would be like is apparent from the fact that he did not provide for the arming of the troops operating the lines of communication, nor did he prepare for German civilian firms to assist in the restoration of the Belgian railways.[28] As it was, one could only trust to finding adequate quantities of food and fodder in Belgium; set up as many formations of *Eisenbahntruppe* as possible, and send them forward as far as possible; and hope for the best.

The Plan modified

On 1 January 1906, Schlieffen was placed on the retired list and his function as chief of the General Staff taken over by Helmut von Moltke Jr. The latter has since been blamed by generations of historians, first for tampering with the Master's design and then for lacking the resolution to carry it out. And indeed, as far as the logistic aspect was concerned, Moltke was much less ready than his predecessor to stake Germany's future on hazy, ill-defined expectations of loot and extraordinary good luck. Scarcely one month after he had taken office, the first realistic study of the supply and transportation problems of the great Plan was written. Its author was the head of the railway section of the General Staff, Lieutenant Colonel Gröner, who was later to turn into the leading exponent of Schlieffen's thought; he cannot therefore be suspected of excessive caution. Nevertheless he concluded that, as it stood then, the Plan stood little chance of success. Gröner did not share Schlieffen's optimistic assumption about the Germans being able to live largely off the country. In his opinion the advance would be far too rapid to allow for a thorough organization of supply feeders, required to sustain a huge army, from Belgium and France. Everything would therefore depend on the regular operation of the railways, and great difficulties were to be expected 'if the railroads are thoroughly destroyed'. Horse-drawn transport, Gröner clearly saw, would not be able to keep up with the advance, so that the moment could be foreseen when 'the Armies would have to halt and let the supply columns catch up'. In this situation, motor transport could be very useful, but Gröner foresaw, correctly as it turned out, that it would be a long time before the German army was able to acquire an adequate supply of this.[29] Clearly, this was not an optimistic forecast. The man who, more than anybody else, was to be responsible for keeping the stream of supplies following behind the right wing armies was very doubtful whether it could be done.

Apart from initiating the first serious study of the Plan's logistic aspects, Moltke felt that the whole subject of supply and subsistence had been neglected by his predecessor. Consequently he instituted, in addition to the normal staff-rides of the General Staff, the so-called '*Mehlreise*' (literally flour-rides) in which subordinates were to be trained in the intricacies of transport and supply. He drove this sensible policy through against considerable opposition, and was to re-emphasize the difficulty of conducting real operations, as distinct from the war games beloved by his predecessor, during the last major exercise he directed shortly before the war.[30] Time and again, Moltke expressed his doubts about the feasibility of the great Plan, indeed about Schlieffen's whole image of war.[31] Given this, it is surprising that he retained the basic outline of the design, not that he modified it so much.

This does not mean that, even from a strictly logistic point of view, the changes introduced were all beneficial. Unlike Schlieffen, who seems to have entertained some illusions on the matter, Moltke did not expect the Dutch tamely to allow the Germans through. Rather than violate their neutrality and take on another opponent, he decided that careful staff work would, after all, make it possible to carry out the Plan without trespassing on Dutch territory. Schlieffen had provided for two Armies consisting of sixteen corps (including seven reserve corps forming a separate echelon) and five cavalry divisions to 'cross the Meuse by five routes below Liège [that is, through the Maastricht Appendix]...and one above it',[32] but Moltke disagreed with this and contrived to march 1. and 2. Armies, forming the hammer-head of the right wing, through the narrow gap between the Dutch frontier and the Ardennes. Whatever the political merits of this decision, the number of roads available to these Armies was thereby cut by half, thus compelling them to march behind each other instead of side by side and imposing a delay of about three days. At the same time, the need to employ the maximum number of railroads made it impossible to narrow down the area of concentration, so that another forty miles were added to the distance to be covered by 1. Army, which was now supposed to carry out two sharp changes of direction, by marching around the Maastricht Appendix instead of through it. Since Kluck's troops were now to enter Belgium from the south-east instead of the north-east, it became much more difficult to include the Belgian Army in the turning movement. There was the danger that it would escape into the great fortress of Antwerp, which was just what happened eventually. Finally, the decision to respect Dutch neutrality meant that another basic tenet of German military doctrine[33] was abandoned. As 2. and 1. Armies defiled through the Liège gap the six corps making up each of them would be reduced to three roads only, which would lead to the formation of huge columns some eighty miles long and, inevitably, congestion and a loss of contact between combat units and their logistic support.[34] Under such circumstances, the whole system of supply was certain to be thrown into disarray. During its advance to the Meuse, 1. Army would have to live off provisions sent ahead to the railways at Bleyberg, Morsnet and Henri-Chapelle.[35]

The question as to how many troops should – and could – be made to operate in Belgium north of the Meuse was also affected by Moltke's decision to leave the Maastricht Appendix alone. Schlieffen, as we have seen, wanted to employ sixteen corps and five cavalry divisions (some of these forces were not yet available in his day) in this area, to be followed by a number of *Landwehr*, or second reserve, formations whose task it was to take over the lines of communication and invest such fortresses as might be left standing in the rear. To keep these forces supplied, Schlieffen apparently counted on having three separate double-tracked railways, including two passing through Dutch territory at Maastricht and Roermond.[36] Now that these could no longer be relied upon, 1. and 2. Armies would have to share the line from Aix-la-Chapelle (Aachen) to Liège, which meant that the maximum number of corps that could operate – in the first instance, at any rate – north of the Meuse went down to twelve.[37] For this reduction of the forces on the extreme right wing – the famous 'Verwasserung' of the Schlieffen Plan – Moltke has been severely taken to task. However, the change was more apparent than real. Firstly, the decision to respect the neutrality of the Netherlands made it unnecessary to allocate any troops to contain the Dutch, whose Army, numbering approximately 90,000 men, was held in some respect by the Germans – more so, indeed, than the Belgian one[38] – and would have tied down at least two corps. Secondly, Schlieffen expected to employ no less than five corps to invest Antwerp,[39] whereas his successor finally made do with only two. Though it is therefore quite true that Moltke's right wing was not as strong as Schlieffen had planned to make it, this loss was more than compensated for by the economies effected in his version of the Plan.

If the merits of Moltke's decision to spare the Netherlands a German invasion and accept the consequent technical complications are open to debate, there is one aspect in which his version of the Plan was definitely superior to his predecessor's. Schlieffen, we have noted, was much concerned with the question of the size of the German wheel through Belgium. From 1897 to 1905 it constantly grew larger until it embraced, first Namur, then Brussels, and finally Dunkirk as well. The marching distances involved were enormous and Moltke, who did not share his predecessor's almost monomaniacal preoccupation with the danger of open flanks, was certainly in the right when he determined that Brussels was as far as the German Army would go before beginning its great turning movement to the south-west. This alteration of the Plan involved the additional complication of compelling 1. and 2. Armies to contract their front and arrange their corps behind each other as they passed through the defile between Brussels and Namur, but this was more than compensated for by the reduction of the distance involved by almost one hundred miles. Carrying out Moltke's 'small' wheel in 1914, the German forces somehow kept going until, albeit literally staggering with fatigue, they reached the Marne some 300 miles from their starting point. Had they tried to brush the Channel in accordance with Schlieffen's prescription, sheer exhaustion would certainly have brought the advance to a halt long before it ever reached the lower Seine.

Logistics during the campaign of the Marne

It can be said that World War I broke out on 1 August 1914, the date on which most European powers ordered general mobilization. Ten days later, the German Army was deployed on the Reich's frontiers according to plan, and the preliminaries of occupying Liège and Luxemburg had been successfully completed. The great wheel across Belgium was now ready to get under way. In this wheel, von Kluck's 1. Army operating on the extreme right wing was destined to play a role of crucial importance. Since it would have to cover the greatest distances at the highest speed, its logistic problems would perforce be the most difficult, reflecting, as it were, those of the Army as a whole in a magnified form. For this reason we shall concentrate our discussion on this force, referring to the others when relevant.

Setting out from its area of concentration around Krefeld and Jülich on 12 August, the Army found itself marching down an inverted funnel which grew progressively narrower as the advance went on. By the time Aix-la-Chapelle was reached, the Army's six infantry corps (its cavalry corps had preceded it through the Liège gap, being subordinated to 2. Army for the purpose) had to share three roads between them, a situation which persisted until they got across the Meuse thirty miles to the west. Already in this early stage of the advance – indeed, even before the frontier into Belgium had been crossed – the heavy (Army) transport companies were falling behind, and were soon to find themselves separated from the units, whose organic supply vehicles they were supposed to replenish, by miles of endlessly-marching troops.[40] Fortunately, the region up to, and including, the Meuse had already been more or less cleared by Bülow's 2. Army. Apart from the occasional straggler or *franc-tireur*, Kluck's troops met with no resistance and were kept supplied directly from the Aix-la-Chapelle–Liège railroad.

Having defiled through Liège, 1. Army changed direction from south-west to north-west and, expanding its front laterally, began racing the Belgian army to Brussels. Though the country had now opened up sufficiently for each corps to have a road of its own, the heavy columns had been left so far behind that they could not catch up with the fighting troops until *after* the retreat from the Marne.[41] As was to be expected, the forces forming Kluck's extreme right were the first to feel the strain, and by 19 August – only three days after crossing the German–Belgian border – they were beginning to fall behind schedule.[42] Consequently, the attempt to 'scoop up' the Belgian army in the enveloping movement, never very promising since Moltke's decision to change the direction of the advance and refrain from violating Dutch neutrality, was doomed to failure.

As 1. Army lost contact with its heavy transport columns during the very first days of the campaign, it quickly became clear that the arrangements made to provide the troops with subsistence were hopelessly inadequate. Captain Bloem's company, forming part of III. Reserve Corps, was typical in that it did not catch a single glimpse of the transportation companies during the entire advance.[43] Fortunately for the Germans, the country they were traversing was rich, and the season of the year favourable. Also the advance had been so rapid that the retreating Belgians often failed either to destroy or evacuate their supply dumps. Thus, III. Reserve Corps, mentioned above, was able to manage without having to draw anything from its organic transport except for some vegetables and coffee. Sharing a road with III. Reserve Corps at the beginning of the march, IX. Corps was so fortunate as to find vast stores of Belgian flour at Liège. Having entered Brussels on 20 August, 1. Army promptly requisitioned enough food to fill the needs of four corps for one day. Again, at Amiens, IV. Reserve Corps found subsistence in considerable quantities.[44] After the battle of Le Cateau, III. Reserve Corps was living well off British loot.[45] So, thirty years of dire warnings, uttered by everybody from the great Moltke downward, about the inability of modern *Millionenheere* to exist in the field turned out to be wrong. Instead, Schlieffen's confident view that it would be possible to more or less fill the bellies of the men from the country was proved correct.

There were, of course, problems. Each army corps, a complete little army in itself, consumed about 130 tons of food and fodder a day,[46] and to find such vast quantities foraging parties had to be sent out over a large area, increasing still further the length of the daily marches. While many items were fairly easy to obtain, bread – the most important single constituent of the soldier's diet – was always in short supply, either because it went stale on the way or because the mobile field-kitchens were not allowed to stay sufficiently long at any one place for baking to be completed. Likewise, the arrangements made to supply fresh meat by purchasing local cattle and driving it along with the Army proved such a failure that the transport allocated for this purpose was soon put to other uses.[47] Finally, the attempt to increase the mobility of the cavalry by depriving them of their own organic subsistence companies was unsuccessful. Instead of happily 'travelling light', the cavalry commanders became unduly fussy about their supplies, racing (and, of course, beating) the infantry to such food and shelter as were to be had, and impeding their own freedom of movement by impressing heavy Belgian peasant-wagons into their service.[48]

On the whole, the men were, therefore, able to live – and sometimes live well – off the country. It was only on odd occasions, especially in the period immediately before and during the battle of the Marne, that it was necessary to resort to the iron rations carried by the soldiers.[49] Since the country tended to become even richer as the advance continued, it is fairly certain that, despite the occasional hungry day, the problem of feeding the troops would not have presented insuperable difficulties even if the battle had gone in Germany's favour.

This, however, does not hold true for the horses' fodder. The German experience in 1914 served to confirm the old wisdom that the animals accompanying an army were very much more difficult to subsist than its men. Years before the War, there had been warning voices raised against placing any reliance on the resources of the country to sustain large masses of cavalry,[50] but both Schlieffen and Moltke had chosen to ignore them. In fact there was little else they could do, for the fodder requirement of the German Army in 1914 was so huge (Kluck alone had 84,000 horses consuming nearly two million pounds per day, enough to fill 924 standard-model fodder wagons) that any attempt to bring it up from base by means of the *Etappen* system would have made the whole campaign utterly impossible. Consequently, the Germans entered the War with little or no arrangements to feed their horses in the field, and were again fortunate in that the season of the year was very favourable. Fodder was frequently found, ready-harvested and neatly stacked, in the fields, and could sometimes be processed on the spot with the help of local machinery.[51] Most of the time, however, it was necessary to feed the horses green corn, causing weakness and sickness that could not be effectively dealt with, as there was no proper field veterinary service.[52] So bad were the arrangements made to feed the horses that some of the artillery teams died very early in the campaign, sometimes even before the border into Belgium had been crossed.[53] Cavalry commanders repeatedly complained to OHL about the shortage of fodder, the reply invariably being a bland exhortation to live off the country even if this meant curtailing the pace of the advance.[54]

The failure to pay sufficient attention to the problem of feeding the horses did, in fact, have its effects very early in the campaign. Already on 11 August one cavalry division, its horses starving and exhausted, had to be taken out of the line. Two days later, an order for all the cavalry forces preceding 1. and 2. Armies to halt and rest for four days had to be issued. In spite of this breathing-space, 2. cavalry division (1. Army) was again brought to a standstill by supply difficulties on 19 August, and by the time the Germans crossed into France all the horsed forces were suffering from exhaustion. On the eve of the battle of the Marne, the German heavy artillery – like the rest, horse-drawn – the one arm in which they did enjoy a definite qualitative advantage, was no longer able to keep up, and the cavalry was incurring unnecessary casualties because the horses were too weak to carry their riders out of danger quickly.[55] By this time, too, one German Army at least was finding that the state of the cavalry seriously interfered with operations. As Moltke himself put it, the army no longer had a single horse capable of dragging itself forward.[56]

If the supply of food could – at the cost of an occasional hungry day – be more or less improvised, and that of fodder ignored until the horses dropped dead, ammunition presented a more serious problem. Precision-made modern arms require their own specific ammunition and spare parts. The days when a Napoleon could simply incorporate the entire Austrian arsenal, lock, stock and barrel, in the armaments of the *Grande Armée* were over. Small arms, machine guns, field artillery, howitzers and heavy artillery all had to be kept supplied, and supplied at a rate that had never been thought possible before the war. Here, again, the horse-drawn heavy columns failed completely to the extent that, instead of fulfilling their proper function, they found themselves used—or rather, not used— as rolling magazines.[57] The entire task of supplying the right-wing with ammunition thus fell to the wholly inadequate number of motor-transport companies available.[58] These, together with miscellaneous requisitioned vehicles and a civilian car-park set up by some enterprising citizens of Aix-la-Chapelle, proved to be of a value out of all proportion to their number when the great test came in 1914.

The problems encountered by the motor transport companies as they struggled to maintain the flow of ammunition are interesting because they are typical of an Army which, though standing on the threshold of a new mechanical age, had not yet adapted either its instruments of control or its thought-processes to the newly-acquired technical means. Though the right wing Armies did have a considerable number of lorries between them, means to guide and supervise the columns were lacking, and the only way to contact convoys on the move was to send out hordes of staff officers to find and intercept them. Furthermore, German intendants at all levels were trained to give ammunition absolute priority over all other items of supply. This order was obeyed very strictly, indeed so strictly that the drivers whose lorries were 1. Army's only effective carrier of ammunition were often unable to get their tanks refilled with petrol.[59] Important as they were, the motor-transport companies were utilized inefficiently, and could not adapt themselves to the rapidly changing tactical situation.

In addition, there were all the usual problems that make the supply of a fast-advancing army such a difficult operation. Motor lorries, which according to regulations were to cover no more than sixty miles a day, six days a week, were in fact driven so hard that sixty per cent of them had broken down by the time the battle of the Marne was fought. As drivers worked round the clock, fatigue was responsible for many accidents. Spare parts, and tyres in particular, were almost impossible to obtain, because of the enormous variety of vehicles in use, and the pressing into service of locally-requisitioned vehicles only made the situation worse.[60] Also, the erratic rate at which ammunition was consumed meant that the advancing motor-columns often found the regimental transport wagons still filled up with unused rounds and were thus unable to unload their cargoes. In such cases, field commanders were tempted to 'hijack' the trucks and use them as rolling magazines. Alternatively, they would be sent back with their loads intact. Either practice would result in the columns doing nothing useful for days on end, but neither could be entirely stamped out despite the issue of strict orders.[61] By 24 August a shortage of ammunition, especially for the artillery, began to make itself felt.[62] Fortunately for 1. Army, consumption fell very sharply after the battle of Le Cateau on 26 August. Had this not been the case, the supply service would, in all probability, have broken down.

Though the distances involved and the sheer magnitude of the task were responsible for most of the difficulties in maintaining the flow of ammunition, some were due to faulty organization, carelessness on the troops' part, or plain bureaucratic mismanagement. Deprived of organic transport which OHL feared would impair their mobility, the cavalry divisions were chronically short of ammunition and formed a constant burden on the army corps so unfortunate as to be responsible for them.[63] Ammunition was often unloaded in quantities greater than those needed, and would then be left lying in the open field.[64] Finally, though each Army controlled its own provisions depôt, the supply of ammunition was centralized in the hands of General Sieger of OHL, who was only willing to relinquish his fast-diminishing reserves at the last possible moment, and would then demand that they be sent forward with all possible speed. This arrangement was clearly unsatisfactory. In the future, wrote Gröner in his diary, it would be necessary to give Army commanders complete control over their own stores of ammunition.[65]

The supply difficulties of 1. Army were aggravated still further by the fact that, immediately before the battle of the Marne, its movements had been extremely erratic, and became even more so during the battle itself. Coming from the north on 26 August, Kluck turned south-west in pursuit of the British Expeditionary Force he had beaten at Le Cateau, then changed the direction of his advance to the south-east on 31 August. Having crossed the Marne, 1. Army's corps had to be wheeled abruptly west across their lines of communication in order to face the French on the Ourq. Finally, after the order for the retreat to the Aisne was given on 9 September, a corps had to be sent east again across the Army's lines of communication in order to avoid losing contact with 2. Army on its left. That the supply of ammunition, and indeed communications in general, did not break down during this confused period should be remembered as a triumph of staff work. As the battle approached its end though, the effect of the marches and countermarches was beginning to tell, and there was considerable confusion and congestion in 1. Army's communications. Driving over to meet Kluck on 9 September, Lieutenant-Colonel Hentsch, on his fateful mission, got entangled in this confusion and had to resort to force in order to find his way out.[66] That the unfavourable impression thus created contributed to his decision to have the Germans retreat to the Aisne can well be imagined.

Nevertheless, in spite of all difficulties, the supply of ammunition to 1. Army did not break down during the battle of the Marne, nor is there any evidence that severe shortages were experienced by the other right wing Armies. Such shortages developed only after the battle was over – the first orders to conserve ammunition went out on 15 September[67] – and then they did not result from any transportation difficulties within or behind the Army, but rose out of the general depletion of the stocks available in Germany.

State of the railroads

Though hundreds of supply companies and tens of thousands of vehicles were crowding the roads of Belgium in August and September 1914, the advance into France could only be sustained if new forward railheads could be opened at a pace comparable to that of the troops. Though no precise details are known, it appears the the Germans had counted on utilizing four distinct lines to supply their five right wing Armies. One of these was to follow Kluck through Liège, Louvain, Brussels and Cambrai, and to provide the logistic support of 1. Army and the right wing of 2. Army. Then there was to be another line running south-west from Liège to Namur, which was to sustain 2. and 3. Armies. 4. and 5. Armies near the pivot of the great wheel were to be fed from two lines passing through Luxembourg and from there on to Libramont–Namur. At a later stage, it was expected to have a line from Metz to Sedan in operation.[68] Since it could not of course be foreseen exactly which lines would be most heavily damaged by demolitions, the scheme was necessarily a general one and depended for its implementation on the rapid repair of blocked track.

As it was, this task proved heavier than expected. The deeper the advance into Belgium and northern France, the more extensive the demolitions became until, midway through September, the 26,000 men of the railway construction companies were no longer able to cope, and had to be supplemented by German civilian firms which alone possessed the capacity to carry out thorough repairs. In the meantime, the Army had to manage as best it could, and this was clearly not good enough. Out of forty-four major *Kunstbauten* blown up or otherwise destroyed in Belgium, only three had been restored by the time of the battle of the Marne.[69] At the same period, only three or four hundred out of the 2,500 miles forming the Belgian network were back in operation. Nor do these figures, revealing as they are, tell the whole story. Even where tracks were found more or less intact, signal and communications gear was usually lacking, having been dismantled either by the Belgians or by the German vanguard itself. The strength of the rails, as well as the length of the crossings and side-tracks, often proved insufficient to take fully-laden German military trains. Already overburdened with work, the unfortunate *Eisenbahntruppe* were compelled to guard the railways against marauding enemy civilians and cavalry raids. Very little Belgian rolling stock was captured in the early stages of the campaign, and when larger quantities were found later it was blocking the rails and had to be evacuated and rearranged.

Over these railways, sometimes precariously repaired[70] and usually stripped of even the most basic equipment, traffic was initially chaotic. No very great performances could be expected of those parts of the network that had been put back in operation, but zone-of-communication authorities, eager to satisfy the demand for supplies, reduced their efficiency still further by rushing through the greatest possible number of trains, regardless of the consequences to the subsequent working of the lines.[71] Impatient field commanders often interfered with the traffic, either 'hijacking' trains destined for other units or putting wagons out of operation by using them as convenient magazines. Since the movements of the right-wing Armies, especially after 30 August, were erratic and subject to unexpected changes of direction, it frequently happened that trains loaded with supplies for certain units were unable to locate them, and subsequently got lost until somebody at headquarters remembered to inquire about their fate.[72] So-called 'wild' wagons, trains loaded with well-meant presents to the troops, were often sent back from railheads unable to receive them, and would then roam the network with little or no control from above.[73] All these were temporary shortcomings that time, experience and strict discipline would cure. By the time they were cured, however, the battle of the Marne had been fought and lost.

The effectiveness of the lines of communication behind each individual Army varied considerably. It was Kluck's force – paradoxically, the one with the longest distance to cover – whose situation was best. Apparently surprised by the direction of the advance, the Belgians had not had the time to demolish the railways in front of him thoroughly. Obstructions of a minor kind – demolished tracks, tunnels blocked by running trains into each other (on one occasion seventeen locomotives were used to this purpose) and the like – were fairly common, but could be repaired with relative ease. Working round the clock, the *Eisenbahntruppe* opened Landen to traffic on 22 August, Louvain two days later, Cambrai on the 30th, and Saint-Quentin on 4 September. Nevertheless, the railway system behind 1. Army was far from satisfactory, as the following table shows:

Date:	*Front-line passing through*:	*Railhead at*:	*Distance*:
22 August	Brussels	Landen	40 miles
24 August	Condé	Louvain	70 miles
26 August	Crèvecoeur	Brussels	80 miles
29 August	Albert–Péronne	Mons	65 miles
30 August	Corbie–Chaulnes–Nesle	Cambrai	40 miles
4 September	Coulommiers–Esternay	Saint-Quentin	85 miles
5 September	Coulommiers–Esternay	Chauny	60 miles

For much, indeed most, of the time Kluck's troops were operating well beyond effective supporting distance from the railheads – a fact which was all the more serious because the above figures represent a minimum which, in many cases, was largely theoretical. The opening to railway traffic of any given point did not automatically mean that the Army's transport companies would henceforward be able to collect all their loads from that point. Rather, as one station after another was reached and then left behind, there was a tendency for stores to accumulate in a series of dumps all along the line. For example, though Saint-Quentin was serving as 1. Army's railhead on the eve of the battle of the

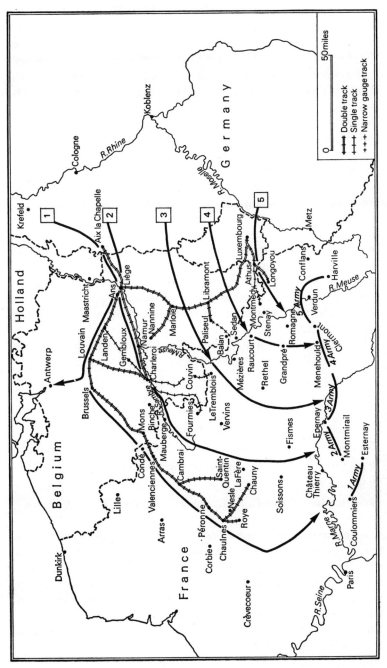

6. German Railway Supply Network in Germany and France, 5 September 1914

Marne, much of Kluck's ammunition was still stored in magazines at Valenciennes or even as far back as Mons.[74] With the *nearest* supply point 85 miles behind the front, difficulties were inevitable.

If the rail support of 1. Army worked more or less in accordance with the original plans, the situation of 2. Army to its left was very different. Advancing south-west along the Sambre, Bülow found his natural line of communication blocked by the fortress of Namur. The town fell on 23 August, but the bridge over the Sambre was so heavily damaged that the section to Charleroi could not be opened to traffic until nine days later. Meanwhile, 2. Army's supplies had to make a big detour from Liège to Landen, and from there southward by single track to Gembloux (23 August), Charleroi (25 August, which was just on time as Bülow had threatened to suspend his advance unless this particular railhead was restored to service)[75] and Fourmies (30 August). The line from Aix-la-Chapelle through Liège to Louvain was therefore heavily burdened by the supplies of two Armies. From 30 August to 2 September, yet another Army – Haussen's Third – was also supplied by the same route, so that each of them was receiving just six trains per day.[76]

Meanwhile, the distances separating 2. Army from its railheads were gradually increasing, as shown in the following table:

Date:	Front-line passing through:	Railhead at:	Distance:
23 August	Binche–Thuin–Namur	Gembloux	22 miles
25 August	Thuin–Gevet	Charleroi	20 miles
30 August	Saint-Quentin–Vervins	Fourmies	30 miles
2 September	Soissons–Fismes	Fourmies	95 miles
4 September	Montmirail–Epernay	Couvin	105 miles

Keeping well within supporting distance during the early part of the campaign, 2. Army rapidly outran its communications at the very time when the opening of the battle of the Marne drastically increased the consumption of ammunition. Likewise, 3. Army was also receiving its supplies from Couvin on 4 September, from where 85 miles had to be covered by road along the line Epernay Châlons sur Marne. However, from Couvin to Le Tremblois there was a narrow-gauge line, described as '*wenig leistungfähig*', which reduced the distance by some fifteen miles.

In addition to this, the railway in question suffered even greater difficulties than those afflicting the network in general. These began at the upper end between Ans and Liège, where the track was so steep that four locomotives had to pull and push each train forward. Further down, Liège formed a bottleneck that was often congested. Each day brought its special problems. On 18 August the transport of heavy siege artillery toward Namur caused the suspension of most other transports, and on 21 August an accident at Liège blocked the line. On 23 August there was another accident, this time at Ans. On the following day supply trains were held up by troop-transports carrying IX. Reserve corps from Schleswig–Holstein. In the last days of the month, communications between Liège and Aix-la-Chapelle – seat of the crucially important *Militäreisenbahndirektion* I – were interrupted. Thus the railway serving as the Germans' main artery of supply in Belgium was in an almost permanent state of crisis.

Further to the east, things were somewhat better. The 4. Army had a double-tracked line reaching Libramont from Luxembourg on 16 August, and a single one to Paliseul a fortnight later. From Paliseul, a narrow-gauge line reached the destroyed bridge over the Meuse at Balan on 1 September. At this point supplies were ferried across the river, then sent by single-track standard-gauge line from Sedan to Raucourt, which was opened to traffic on 4 September. Consequently, distances between the railheads and the front were as follows:

Date:	Front-line passing through:	Railhead at:	Distance:
30 August	Rethel–Stenay	Paliseul	70
1 September	Grandpré	Balan	50
4 September	Menehould	Raucourt	60

Though the distances to be covered by road were not excessive, compared with those behind the right wing Armies, 4. Army depended on a rather complicated logistic chain that limited its supplies to some 300 tons a day, mainly ammunition.[77] Compared with this, the situation of 5. Army near the pivot of Schlieffen's gigantic *Klaptur*, was best of all:

Date:	Front-line passing through:	Railhead at:	Distance:
25 August	Montmédy–Conflans	Veriton	20
30 August	Stenay–Romagne	Montmédy	15
4 September	Clermont–Harville on either side of Verdun	Montmédy	45 and 40 respectively

At Montmédy, a tunnel had been destroyed and remained out of service until the end of October.

Overburdened and at times close to collapse, as they were, there is no direct evidence to suggest that the failure of the railways to carry sufficient traffic, or to keep up with the pace of the advance, played any significant part in the German defeat on the Marne. Had that battle gone in their favour, however, there is every reason to believe that the state of the railway network would have prevented the Germans from following up their victory and penetrating further into France. We have already noted that the scope and intensity of demolitions tended to increase with the duration of the campaign. Even Kluck, who had enjoyed incredible good fortune in finding the line to Chauny and Saint-Quentin intact, was meeting with serious obstacles early in September. Had there been no retreat to the Aisne, these would have taken him a long time to repair.[78] Of the other German forces, 5. Army could probably have continued its march for a very few days until it got out of range of Montmédy. For 2., 3. and 4. Armies any further advance was clearly impossible.

Strength and reinforcement of the right wing

Throughout his years as chief of the General Staff, Schlieffen had repeated doubts as to whether the forces available would be sufficient to carry out the great Plan.[79] German utilization of national resources, especially manpower, was nowhere as comprehensive as that of France. Though her population exceeded that of her rival by some two thirds, she found it hard even to obtain parity in the forces deployed against her. Schlieffen had 'solved' this problem in his usual facile manner by including in his *Aufmarsch* a number of nonexistent corps. Though these additional forces were available in 1914, transport limitations prevented their use on the right wing and they were sent to Lorraine instead.

Although Schlieffen recognized that, 'like all previous conquerors', the German armies would incur very heavy losses during the advance through Belgium and France, he did not draw the logical conclusion of making sufficient reserves available. Nor was the plan of campaign suited to contribute towards a solution, involving as it did an extension of the German front from 140 miles at the beginning of the march to about 190 by the time Brussels was reached – not to mention Schlieffen's fantastic vision of a shoulder-to-shoulder advance along the 350 miles from Verdun to Dunkirk, for which his forces were totally inadequate. Even along the far shorter front of 1914, the German troops were not sufficiently dense on the ground. Gaps between the Armies were constantly opening up, the result being that Haussen's 3. Army in the centre was continually changing direction as he tried to answer his neighbours' calls for flank protection.[80] As it was, the German defeat on the Marne was the direct outcome of a thirty-mile gap between 1. and 2. Armies.

To make matters worse, events in August and September 1914 fully justified Schlieffen's fears about the wastage involved in a rapid advance through enemy territory. No complete statistics are available, but evidence compiled from a variety of regimental histories and personal diaries show that, on arriving at the Marne, many German battalions were reduced to half their effectiveness.[81] Losses incurred in combat and sickness, and the need to guard impossibly long lines of communication against a hostile population all contributed to this. Also, there was sheer exhaustion among the men. Schlieffen's Plan demanded very great exertions of the troops of the right wing, some of whom had been covering distances of between 20 and 25 miles a day for many days. On 4 September, the commanders of 1. and 3. Armies both reported that their units were fast approaching the point of collapse.[82] Back at OHL, Moltke was equally well aware of the situation, and seems to have wanted to give all the right wing Armies a day of rest on the 5th. Events made this impossible, and in fact it was only Haussen's troops who did enjoy a breathing-pause. After the campaign was over, and he was no longer chief of the General Staff, Moltke appears to have acknowledged that the uninterrupted rush from the frontier to the Marne had been a mistake.[83]

In so far as he considered providing any reserves at all – he had once written that 'the best reserves are an uninterrupted stream of bullets'[84] – Schlieffen had planned to bring up two corps from his left wing in Lorraine. In 1914, adequate railways were available for this purpose, as was a suitable quantity of rolling-stock, made ready for the purpose from the tenth day of mobilization onward.[85] By the evening of 23 August, the French attack in Lorraine had been delivered and bloodily repulsed; nothing now prevented the two corps from being transported to Belgium as scheduled. In the event, Moltke did not carry out this part of the Plan, preferring instead to go over to the offensive on his left wing also. The reasons for his decision to carry out this so-called '*Extratour*' in Lorraine have been discussed elsewhere and need not concern us here.[86] All we are interested in is whether, from the point of view of transport and supply, it would have been possible to bring up additional forces to the right wing in time for them to intervene in the battle of the Marne.

The speed at which troops can be moved by rail depends on a great many factors, including the number and quality of lines available, the state and quantity of the rolling stock, and, perhaps most important, the number and location of stations available for loading and unloading the trains. Except for the fact that four well-developed railways were available, we have no precise data about any of these, and will therefore have to confine ourselves to a few very elementary calculations.

In 1914, it took 240 trains, with fifty wagons each, to transport two German army corps. Assuming that this many trains were, in fact, available, that all four lines could have been cleared of all other traffic to carry sixty military trains per day, and that there were enough quays to allow the simultaneous loading and unloading of all these trains within short distances from the troops, under these ideal conditions, it should have been possible to cover the 150 miles from the Metz–Diedenhoffen area to Aix-la-Chapelle in about four days, en- and disentraining times included. The movement would thus have been completed on the evening of 27 August.

Apart from the obviously impossible demands made above, this calculation assumes a clockwork efficiency in carrying out all movements. What is more, it does not take account of the fact that 90,000 troops and all their supplies would have had to cover a considerable area in order to find railway stations capable of handling them. Nevertheless, assuming that they could have left Aix-la-Chapelle on 28 August, the two corps would have had just 13 days to march the 300 miles to the Marne and arrive at the *end* of the battle there on 9 September. Even if they could have sustained a pace of twenty miles per day, they would have arrived too late.

At this time it was impossible to transport troops forward from Aix-la-Chapelle by rail. Only one double-tracked line was available to 1. and 2. Armies, and this could handle only some 24 trains per day, of which 4 were needed for the operation of the railways themselves.[87] To bring up the fighting elements of two corps, 120 trains were required. Even assuming that one third of the normal traffic along the line had been stopped, the two corps would not have arrived until *after* the retreat to the Aisne had been completed – this, at a time when Gröner was repeatedly warning the Armies dependent on the line to reduce their demand for supplies to the indispensable minimum.[88]

It has been suggested that, instead of marching on foot, the two corps might have used motor-lorries to travel at '100km. a day' to the threatened right wing.[89] To carry the combat units involved, however, no less than 18,000 vehicles would have been needed,[90] and there were only 4,000 available to the German Army in 1914. Such an idea was obviously impracticable.

There remains the question of whether it would have been logistically possible to employ additional army units – up to 10 corps, according to one authority[91] – on the right wing, had the Germans employed their national resources to set up such units. Given the need for 1. and 2. Armies to pass through two defiles, first at Aix-la-Chapelle–Liège and then between Brussels and Namur, it is certain that road-space would not have allowed more troops to be employed north of the Meuse than was actually the case.[92] It has, however, been suggested that another Army – numbering, perhaps, four corps – could have followed in Kluck's right rear and, by fending off the threat to his flank at the Marne, turned the battle there into a German victory.[93] As the authors of this proposal admit, it would not have been possible to feed these forces from the already overburdened railways. However, they would presumably have expended very little ammunition and could have been made to live off the country. This argument overlooks the fact that the country in question had already been traversed by Kluck's troops, and that supplies, especially fodder, would therefore have been difficult to obtain. Assuming that it would have been necessary to bring up only half of the additional Army's needs from base, it is possible to calculate that some 500 three-ton lorries would have been required.[94] This was more than were available to all five German right wing Armies in August 1914.

Conclusions

Any attempt to give a definite answer to the question whether the Schlieffen Plan was logistically feasible is bound to suffer from a lack of information. Thus, in spite of the enormous literature on the subject, we have no precise data about a great many vital factors, including the consumption of subsistence and ammunition at various times and places during the campaign, the number and loads of trains travelling over the Belgian railroads, the exact number, state and location of the railway stations used, statistics about supplies reaching the troops in the field, and so on.

In so far as it is possible to follow his thought in any detail, Schlieffen does not appear to have devoted much attention to logistics when he evolved his great Plan. He well understood the difficulties likely to be encountered, but made no systematic effort to solve them. Had he done so, he might well have reached the conclusion that the operation was impracticable.

Though not primarily interested in transport and supply, Schlieffen was able to foretell the pattern of railway demolitions in Belgium with uncanny accuracy, in some cases going so far as to name the very installations whose destruction would cause the greatest difficulties in 1914. His prediction that it would be possible to make the Army live off the country proved largely correct, despite the scepticism of practically every well-informed military writer of the day. Had Schlieffen been wrong on this point, the entire campaign would have floundered very soon after its start.

Moltke did much to improve the logistic side of the Plan. Under his direction, the problem was seriously studied for the first time and officers trained in the 'technics' of warfare which, in Wilhelm II's Army, were apt to be disregarded or looked down upon. It was to him that the Army owed the introduction of motor-transport companies without which, again, the operation would have been utterly impossible. He did, it is true, make a number of changes in the Plan. From an exclusively logistic point of view, some of these were beneficial, but most were harmful. Nevertheless, taking his period of office as a whole, he probably did more to improve the Plan than to damage its prospects.

As it was, the Germans enjoyed a success beyond the limitations of the Plan. In 1914, marching distances far exceeded anything that had been thought possible in peacetime.[95] The country was rich, the season of the year most favourable. Though railway demolitions generally proved more serious than expected, they were much less so at the place where they counted most, i.e. behind 1. and 2. Armies. While it is not clear what quantities of supplies were carried forward over the network, the heavily-damaged railways of 1914, though difficult to operate, proved more or less capable of feeding the army. Even at the time of the greatest pressure, when no less than three Armies were dependent on one line, the six trains reaching each of them daily carried sufficient supplies to meet the most urgent requirements.[96] At least, there is no evidence to the contrary.

Though the railways were able to handle the requisite quantities of supplies, it proved impossible to advance the railheads sufficiently fast to keep within supporting distance of the army. By the time of the battle of the Marne, *all* the German Armies save one had gone well beyond that distance. Only for Kluck was there any prospect of a rapid reconstruction had the battle been won, and even he would probably have met with the greatest difficulties. It would have been impossible to keep the other Armies supplied at all.

In spite of a thorough reorganization carried out in 1908, the second tier of the German supply system – the heavy transport companies operating in the zone of communications – proved a complete failure.[97] Their task admittedly had been made heavier by Moltke's provision for more than one corps per road, a measure which threw the entire organization out of gear before the campaign even started, and which could not be corrected thereafter. Given the relative speed of marching columns and horse-drawn vehicles, however, the troops would have outdistanced their transport anyway. For much of the German Army in 1914, one crucial link in the *Etappen* system proved simply irrelevant. Its place had to be taken by motor transport columns, and it was thanks to the enormous efforts they made that the advance got as far as it did.

Although there were temporary shortages and hungry days, supply difficulties were not responsible for the German defeat on the Marne. Food was obtained from the country, horses went unfed until they died, and ammunition somehow did arrive from the rear in more or less adequate quantities. In August and September 1914, no German unit lost any engagement because of material shortages.

Had the battle gone in Germany's favour, however, there is every reason to believe that the advance would have petered out. The prime factors would have been the inability of the railheads to keep up with the advance, the lack of fodder, and sheer exhaustion. In this sense, but in no other, it is true to say that the Schlieffen Plan was logistically impracticable.

As it was, the entire 'technical' side of the great design was characterized, not by the thorough and methodical planning commonly associated with the German General Staff, but by an ostrich-like refusal on Schlieffen's part to face even those problems which, after forty years of peace, could be foreseen. Moltke did much to improve matters in this respect, but in the final account it was furious improvisation, not reliance on carefully-made preparations, that enabled the Army to reach as far as it did. As the officer responsible for the communications zone behind 1. Army was to put it:

> The hurry and stress of the first few months were so great
> that. . .the whole doctrine of supply had been somewhat
> upset by the rapid advance. The service of supply could
> not, therefore, adhere closely to previously accepted
> principles. For this reason it did no special harm when we
> were issued. . .all the various regulations dealing with the
> service of supply etc. . .These regulations were carried
> along, nailed up in a packing box; but, as far as I know,
> the box was never opened until the Army withdrew behind
> the Aisne.[98]

That the Army achieved as much as it did, at a time when the standing orders could only be said to have caused no actual harm, is remarkable indeed. Critics of the advance would do well to keep this in mind.

The Schlieffen Plan proved to be the last of its kind in several ways. As Liddell Hart has written, manoeuvres of the scope and boldness planned by Schlieffen had been possible in Napoleon's day, and motor transport would make them feasible again in the next generation.[99] Meanwhile, the sheer size and weight of the

German Army in 1914 proved wholly out of proportion to the means of tactical transportation at its disposal. This was true even though the really great increase in consumption came only after the campaign of the Marne was over. In 1914, a British division required just twenty-seven wagon-loads of supplies of all classes per day. Two years later, daily consumption of current supplies – food, fodder and the like – still stood at twenty wagons, but the number needed to carry the material of combat, especially ammunition, had risen to about thirty.[100] After 1914 subsistence for men and horses was to form only a fraction, and usually a small fraction, of the total supplies needed by armies in the field. For this very reason, it was no longer possible to meet a good part of their needs on the spot. That the old modes of transport were inadequate to handle the demands of modern war, is demonstrated by the permanently-fixed lines of trenches that were a hallmark of World War I.

As happens so often in history, the lessons of this experience were misunderstood. The German campaign of the Marne, and especially the operation of the *Feldeisenbahnwesen*, were analysed time and again between the Wars. It was shown that, had a few tunnels been blown up along Kluck's route, the entire campaign would have been utterly impossible. In 1940, the Belgians thought they had learnt their lesson by mining every one of their major railway-installations and preparing them for instant demolition. By then, however, the face of war had changed once more.

5

Russian roulette

Problems of the semi-motorized army

When Hitler came to power in January 1933 he brought with him a firm, if ill-defined, commitment to the modernization and mechanization of German life which, however useful and even necessary it may have been on economic and strategic grounds, cannot be adequately understood on the basis of utilitarianism alone. The National Socialist Party had always taken a great interest in motor-cars and made much use of them in its rallies, parades and demonstrations. It even incorporated a special drivers' corps, the NSKK. Hitler himself loved cars, had a surprisingly good understanding of their technology and construction, and showed a lively interest in their engines, the men who drove them and the roads built for their use, far beyond the interest that a head of state could normally be expected to take in what was, after all, only one element in his country's system of transportation, and that not the most important. White-coloured *Autobahnen* and bug-shaped Volkswagens stood out as National Socialist showpieces *par excellence*. Far from being merely an instrument in the hands of the Third Reich, the motor-car stood, in more ways than one, as its symbol.

From the point of view of the German Army, especially from that of the various departments responsible for supply and transportation, this state of affairs was a mixed blessing. The problem was mainly that Hitler, in this field as in all others, took no interest in administrative detail, nor did he have the patience to carry through the long-term projects which alone might eventually have provided him with a well-balanced motorized army. Instead he had an eye for the spectacular and wanted quick results hence his strong tendency to concentrate on the two ends of the scale, the decorative-representational on the one hand and the tactical on the other. After a few years of National Socialist domination the results of this policy could be seen almost daily in countless parades and military tattoos. Along with the columns of vehicles carrying Nazi dignitaries there came armoured and motorized units with their serried ranks of *Panzers* and other fighting vehicles. These were glittering affairs, yet behind the glamour lurked problems, some of which were due to mistaken policies, but others to difficulties so fundamental as to render the entire course on which the Wehrmacht was now embarked highly questionable.

Perhaps the most important issue was the role of roads *vis à vis* railroads. During World War I, the latter had enabled Germany to make full use of her internal lines and thus resist the combined resources of almost the entire world. The recognition of this achievement subsequently served to transform Gröner from head of Schlieffen's railway department into Minister of War of the Weimar Republic. The railways were too clumsy to sustain mobile operations in the field, however, as was proved in 1914 and thereafter in virtually every offensive that was launched on the Western front by both sides throughout the war. Even where a tactical breakthrough was achieved, supplies could not follow. In staking their future success on the tactics of the armoured, self-propelled fighting vehicle, therefore, Hitler and his generals were obliged to cast around for a more flexible logistic instrument, which could only be found in the motor truck.

If motorization of the army's supply service could rightly be regarded as important, indeed indispensable, in securing its future success in the field, the strategic benefits appeared far more dubious. Given the technological conditions of 1939, no less than 1,600 lorries were needed to equal the capacity of just one double-tracked railway line. What is more their greater consumption of just about everything (fuel, personnel, spare parts, maintenance) in relation to payload meant that the railway retained its superiority at distances of over 200 miles. It followed that, while motorization was essential for operational and tactical purposes, its effect on strategy would be limited. Also, however great the effort, there was little chance that motor vehicles would relieve, much less replace, trains as Germany's main form of transportation in the foreseeable future.[1]

As it was, Hitler's decision to motorize his army led him to fall between two stools. The outlay involved meant that the railways suffered comparative neglect, leading to a decline in the total quantity of locomotives and rolling-stock available between 1914 and 1939.[2] At the same time, Germany's automobile industry was insufficiently developed to meet the needs of the new army in addition to civilian requirements. On 1 September 1939 there were just under a million four-wheeled motor vehicles of all descriptions on Germany's roads, a proportion of 1:70 per head of the population compared with 1:10 in the United States. Finally, motorization required rubber and oil, which Germany did not have, instead of coal and steel, which she did. Even before World War II broke out, voices were raised to question the wisdom of a policy relying on raw materials that had to be imported. Despite synthetic manufacture and heroic improvisation, difficulties in obtaining both rubber and oil created problems throughout the war.[3] Thus, although motorization may have been the only way out of the tactical impasse of 1914–18, logistically its advantages were doubtful.

Between 1933, when Hitler embarked on re-armament, and 1939, the capacity of Germany's motor industry proved wholly inadequate to equip the army on the requisite scale. Of 103 divisions available on the eve of the war, just 16 armoured, motorized and 'light' formations were fully motorized, and thus to some extent independent of the railways for both tactical and strategic movement. The rest all marched on foot, and, though a complement of 942 motor vehicles (excluding motor cycles) was the authorized establishment of each infantry division, the bulk of their supplies was carried on 1,200 horse-drawn wagons. Even worse, all these vehicles were organic to their units and earmarked mainly for work inside the zone of operations. To bridge the distance from the depôts to the railheads, only three motor transport regiments (known as *Grosstransportraum*, in contrast to the *Kleinkolonnenraum* of the troops) were available for the whole army, having between them some 9,000 men, 6,600 vehicles (of which twenty per cent were expected to be undergoing repair at any given moment) and a capacity of 19,500 tons.[4] Since not all the Wehrmacht's units were ever likely to see action at one and the same time there is no point in comparing this with figures of consumption. One may, however, note that the Allies in 1944 used no less than 69,400 tons of motor transport to support forty-seven divisions in France, but nevertheless suffered from a grave shortage.

Since vehicles were so very hard to obtain, it was necessary to take a large proportion of them straight from the civilian economy. This resulted in an impossibly large number of types, all of which had to be kept supplied with spare parts at a rate that was greatly accelerated by the demands of war. So heavy were these demands that, during the early years of the war, it proved impossible to maintain even the modest degree of motorization already achieved. In the winter of 1939–40, and again in that of 1940–1, it was necessary partly to demotorize units and services. This was in spite of the fact that, by the latter date, no less than eighty-eight German divisions – some forty per cent of the total – were equipped with captured French material.[5]

Another problem resulting from the fact that the Wehrmacht was only partly motorized was a lack of homogeneity. Instead of being thinly spread over the entire army, Germany's military motor vehicles were concentrated among a small number of units. In effect, this meant that there were two separate forces, one fast and mobile and the other slow and plodding. Coordinating these two heterogeneous parts was difficult as the former outran the latter and, as they were left behind, it was imperative to exercise the strictest control over the movements of the infantry formations in order to prevent them from obstructing the supply convoys of the all-important armoured spearheads. It was a nice problem of selecting roads, calculating the time it would take to move along them, and maintaining traffic discipline – in short, of logistics.

If the equipment of the Wehrmacht's logistic apparatus left something to be desired, so did its organization. In wartime all transport by rail or inland waterways – both civilian and military – was under General Gercke, whose post was that of *Chef des Transportwesens* at the Army High Command (OKH), while at the same time serving as *Wehrmachttransportchef* at the Armed Forces High Command (OKW). Motor transport in the zone of communications, however, was controlled not by him but by General Wagner, quartermaster-general at OKH, who, of course, was also responsible for the supplies which were to be transported. In this way logistic support for the army was split between two authorities, one of whom controlled both ends of the pipeline while the other governed its central section. Worse still, Gercke's authority did not extend to either the navy or the air force. His function with regard to them was a purely coordinating one, and the best he could do was to ask them to put transport-capacity at his disposal.[6]

Even though they were short in duration and conducted over relatively small distances the campaigns of 1939 and 1940 already revealed every one of these problems, and more. During the operation against Poland, destruction of railways by both sides was so heavy that the logistic system was saved from collapse only by the speedy Polish surrender,[7] while the appalling conditions on the roads led to losses of up to fifty per cent of the motor companies' vehicles. To make good these losses proved impossible, for the 1,000 trucks allocated to the land army each quarter out of new production, were insufficient even to replace those lost through normal wear and tear.[8] Faced with a fast-diminishing vehicle park, OKH in January 1940 was driven to reduce the number of motorized supply columns in each infantry division by a half, replacing the rest with horse-drawn wagons, but even this drastic measure could only bring the units to ninety per cent of establishment when the next campaign was started. The army that was to shake the West was a scavenger. Its hopes for victory rested in part on the capture and utilization of French, Dutch and Belgian vehicles.

During the campaign in the West, indeed, the incomplete state of motorization went far to dictate not merely the pace of the advance, but even its form. As the army consisted of two heterogeneous parts, it was initially planned to have the opening attacks carried out by the infantry divisions, thus husbanding the 'fast' units for breakthrough and exploitation. This was in fact standard German practice which they were later to employ in the Balkans as well as against Russia. In the case of France, however, it was feared that the few roads traversing the Ardennes would be blocked by the infantry divisions in their *rangierten Angriff* (setpiece attack) against the river Meuse, and it was therefore decided to carry out the initial offensive by means of armoured and motorized forces.[9] In the event, however, the armoured spearheads moved too fast – Guderian reached the Meuse in three days, instead of five as planned – so that a gap opened up between them and the infantry, a gap which caused Hitler to fear for his left flank and ultimately contributed to his decision to halt the tanks short of Dunkirk .

Once the first part of the campaign had ended with the encirclement and liquidation of the Anglo–French forces in Belgium, the Germans were faced with the problem of resupplying their units in order to continue the advance against the Weygand line. For this purpose it was intended to set up a forward supply base in the Brussels–Lille–Valenciennes–Charleroi–Namur area, a big job made all the more difficult because destruction of railway bridges, tunnels and viaducts was quite as heavy as in 1914. Around 20 May Wagner therefore rang up the Reich Minister for Transport and demanded that 'all the lorries of Germany' be put at his disposal immediately. They were to proceed to Aix-la-Chapelle, where an officer was waiting to form them into columns and to furnish the drivers with 'German Wehrmacht' armbands. A special staff charged with supervising the establishment of a new supply basis in Belgium was sent to Brussels and reported ready to start work on 22 May. On that day a total of 12,000 tons (including 2,000 from Holland) of motor transport were loaded and sent on their way. At the same time Wagner also made use of such trains as could reach Antwerp (fifteen per day as from 24 May onward) and water transport from Duisburg to Brussels. Such were his efforts that the Germans were able to resume their offensive on 5 June.[10]

Though the campaign against France lasted for only six weeks and ended in one of the greatest victories of all history, the army's fundamental logistic weakness did not escape Hitler's attention. Though consumption of many items – especially ammunition – had been moderate, the difficulty of supplying the armoured spearheads during their rapid advance was considerable, and, but for the use of aircraft and the opportune capture of large quantities of fuel, might have brought their attack to a halt. Also the great mass of footslogging infantry had turned out to be scarcely more mobile than twenty-five or even seventy years previously. Lack of marching discipline led to horse-drawn and motor transport sharing the same roads, which consequently became congested and sometimes even blocked. Though demolished railways were repaired with great speed, the *Eisenbahntruppe* were too few in number to work them efficiently, and the quality of the personnel supplied by the civilian *Reichsbahn* for this purpose left something to be desired. Even as the campaign was going on, therefore the Führer ordered the army's supply system to be completely reorganized.[11] In the event, this proved inadequate, and no amount of jugglery was able to make up for Germany's basic weaknesses, such as an insufficiently developed motor industry and an insecure supply of fuel.

Planning for 'Barbarossa'

Among the many thousands of books that have been written about Hitler's Russian adventure, there is probably not one that does not at some point attribute the Wehrmacht's failure, in part at least, to logistic factors – primarily the difficulties resulting from distance and bad roads. However, no detailed logistic study of this campaign, the largest ever conducted on land, has been undertaken so far.

The exact date of Hitler's 'decision' to take on the Soviet Union has been the subject of much controversy among scholars. There is no doubt, though, that the first detailed studies of the problem – whether one cares to regard them as 'operational' or 'contingency' plans – were made in August 1940, when work began simultaneously at both OKH and OKW. In both studies logistic factors naturally loomed large, but, interestingly, the emphasis was very different right from the beginning.

At OKH, the early planning for what was to grow into operation 'Barbarossa' was the task of General Marcks, chief of staff of 18. Army. Marcks was interested mainly in the Russian road-network, the problem consisting in that, whereas the districts north of the Pripet Marshes were better suited for strategic movement on account of the larger number of roads, the country south of them was deemed more favourable for combat. An advance south of the Marshes would lead the Germans into the Ukraine which, though perfect 'tank country', had only one good west to east road via Kiev. On the other hand, though the number of roads north of the Marshes was greater, a thrust in this direction would lead the Wehrmacht into the forests of Belorussia and thus confine the advance to a few widely-separated axes between which little or no contact was possible. Faced with this dilemma, Marcks hesitated, seemingly unable to make up his mind. Finally, he decided to have the best of both worlds, the tactical and the strategic, by recommending that the attack be launched in equal strength along both sides of the Marshes, its objectives being to reach both Moscow and Kiev at the same time.

By contrast, Marcks' opposite number at OKW, Colonel von Lossberg, was interested less in the roads than in the railroads. Realizing from the outset that only the latter could be relied upon to keep the operation supplied, once it got away from the immediate vicinity of the borders of German-controlled Poland and moved into the infinite spaces of Russia, he very properly con-

cluded that the offensive ought to be launched where the railway
lines were best and most numerous, i.e. a thrust straddling both
sides of the great highway from Warsaw to Moscow. A similar
conclusion followed from the need to deploy the forces prior to
the operation, a huge undertaking that could be accomplished
only by means of the railways. Lossberg was not unaware of the
advantages of the 'southern' approach, but it was typical of the
wider view taken at OKW that he saw the main advantage of an
attack in this direction as lying in the shorter distances from the
Rumanian, and, 'later' the east-Galician oilfields. Unlike Marcks,
who was impressed above all by the openness of the Ukraine and
its suitability to armoured warfare, Lossberg also foresaw that
problems would be caused in this area when rain turned the
ground into a morass. However, all these were secondary considerations. It being assumed that, tactically and operationally,
the Germans were superior to their opponents, the main problem
was to keep the operation supplied, and since this could only be
done by means of the railways, along the railways it must go.

As it was, none of these factors proved decisive to Hitler. Adding an economic and an ideological dimension to the various tactical, strategic and logistic considerations advanced by his generals,
he decided first of all that a drive into the Ukraine was indispensable if Germany was to secure grain and, later, oil from the
Caucasus. At the same time, the seizure of Leningrad as the
centre of the Bolshevik *Weltanschauung* was deemed essential.
The result was a double offensive along diverging lines thousands
of miles apart, a situation complicated still further by the fact
that the chief of the Army General Staff, General Halder, wanted
an attack on Moscow and, in deploying his forces, made his dispositions accordingly.[12] The resulting 'Directive No. 21' in which
Hitler laid down his basic orders for the operation was therefore
a rambling and confused document that provided for an advance
by three Army Groups along separate axes towards Kiev, Moscow
and Leningrad up to the general line Dvina–Smolensk–Dnieper.

Returning to the logistic aspect, the problems involved were of
a magnitude far greater than anything experienced before and,
in some ways, since. The number of troops involved, almost three
and a half million, was more than five times that which Napoleon
took across the Niemen in 1812, and this enormous mass, with its
hundreds of thousands of horses and vehicles, had to be marched,
and supported while marching, towards objectives which, from

north to south, lay 600, 700 and 900 miles away from the base of departure. All this was in a country where roads were known to be few and bad, and where every single yard of railroad had to be converted to standard gauge before it could be used; a country, moreover, which was presently furnishing the Reich with considerable quantities of vital strategic materials, from oil to rubber, the supply of which would of course cease the moment operations were launched.

How to secure sufficient stocks of these materials was, in fact, one of the principal concerns of the men who planned the Russian campaign. As always, the first difficulty lay in determining just how much would be needed, but this depended on many factors that could not possibly be foreseen, including the speed with which operations would unfold (which would influence fuel requirements), enemy resistance (which would determine the amount of ammunition needed), and, of course, the duration of the campaign. As it was, the staffs responsible made a number of highly optimistic assumptions: for example, that the expenditure of ammunition would be similar to that of the Western campaign, and that the Russians would be defeated west of the line Dvina–Smolensk–Dnieper. Even so, however, it was still difficult to obtain some items. Tyres, for example, were in such short supply that for some types of vehicles their replacement by steel-shod wheels had to be considered, and production of rubber soles was brought to a standstill. Even though current consumption of fuel was cut back to a level below that which the army regarded as the indispensable minimum, a reserve of no more than three months' consumption (in the case of diesel oil, only one month) could be built up. With the operation due to start in June, a certain shortage of fuel was expected as early as July, though the situation would hopefully be improved thereafter by deliveries from the Rumanian oilfields straight to the army in Russia.[13] Nor was there much hope of utilizing captured fuel, for Russian petrol had a low octane content, and could only be used by German vehicles after the addition of benzol in specially-constructed installations.[14] Since preparations for 'Barbarossa' involved a great expansion in the size of the army (from 9 armoured divisions to 19, and from a total of 120 divisions of all kinds to 180, later revised to 207) spare parts for existing formations were almost impossible to obtain. This problem was aggravated still further by the fact that the Germans in Russia used no less than 2,000 different types of vehicles, the spare parts required by those in the area of Army Group Centre alone numbering well over one million.[15] A further problem lay in the estimation of the consumption of ammunition. An estimate was finally reached in terms of the quantities that could be carried rather than those that would be required. However, this meant that, far from establishing a reserve for twelve months' fighting, as demanded by the commander of the replacement army, the Germans crossed into Russia with only 2–3 *Ausstattungen* (basic loads) plus an unspecified reserve for 20 divisions. It might be expected that these shortages would have induced the German leadership to reconsider the rationale of the whole campaign. Instead of so doing, however, they managed to convince themselves that what they had originally estimated would take five months to achieve could, in fact, be achieved in four, or even in one.[16] The German general staff seemed to have abandoned rational thought at this point. Rather than cutting down their goals to suit their limited means, they persuaded themselves that their original calculations were overcautious, and that the goals could be achieved more easily than they first estimated.

Overshadowing the above problems, however, was the difficulty of obtaining sufficient transport to support the operation. On this point the Germans were facing a fundamental dilemma; the size of the operation was such that it could only be maintained by means of the railways, but this was made impossible because the Russian railways did not conform to the German gauge. To wait until such time as it would be possible to convert them was out of the question, for this would have allowed the Russians to retreat into the depth of their endless country and thus deprive the Wehrmacht of its one chance of victory. It followed that, as was the case in the West a year previously, everything depended on an adequate supply of motor trucks, but these were still in as short a supply as ever. By 'reorganizing' the army's rear services so as to take away their vehicles, purchasing lorries in Switzerland, and replacing those of the civilian economy with captured French vehicles, the OKH succeeded in more or less meeting its own very modest requirement for an average of 20,000 tons *Grosstransportraum* behind each of the three army groups. However, this left no reserves whatsoever, and such was the shortage that seventy-five infantry divisions had to be given two hundred *panje* wagons (a type of peasant cart) each.[17]

Exactly how deep into Russia these lorries would have enabled
the army to penetrate before it was compelled to halt and build
up a new supply basis is almost impossible to say, owing to the
widely differing figures of fuel-consumption and ammunition-
expenditure. It was generally recognized, however, that the
Russians would have to be defeated within the first 500 km (300
miles) if they were to be defeated at all, and the plans made at
OKH did in fact start from the premise that the distance from the
frontier to Smolensk could be covered in one mighty leap, to be
followed by a pause that would allow the railways to catch up.[18]
Assuming – and this is a very optimistic assumption indeed – that
the lorry columns were capable of covering 600 miles in both
directions within the space of six days, loading and unloading
times included,[19] the daily tonnage that could be delivered to the
army with its 144 divisions would amount to $60,000:6 = 10,000$
tons, giving an average per division of just 70 tons a day, of
which well over a third would consist of rations. To consider the
problem differently, one might estimate the requirements of the
33 'fast' divisions (plus their supporting troops, headquarters etc)
at 300 tons a day each, which would mean that, at a distance of
300 miles from the starting point, all the army's *Grosstransportraum* put together would be barely sufficient to supply them
alone, leaving absolutely nothing for the remaining 111 divisions.
There is no detailed evidence of this, and these figures rest on
nothing but our own elementary calculations. Nevertheless, they
do appear to show that the hopes of OKH were too sanguine and
that the army would meet with supply difficulties long before all
its divisions penetrated 300 miles into Russia.

In fact, however, OKH did not intend to supply its armoured
and motorized spearheads by means of lorry columns shuttling
between them and the frontier. To do so would be impossible,
even if enough lorries had been available, for the German army,
in this campaign as in the previous ones, marched in two heterogeneous bodies one behind the other. Sending the lorries supplying the forward units back through the advancing infantry
divisions in their rear would inevitably lead to further congestion
on roads that were already badly congested. Rather, it was a
question of making the armoured spearheads independent of
base-supply during the first stages of the operation. For this
purpose, the normal fuel-carrying capacity of each armoured and
motorized division, amounting to some 430 tons,[20] was supple-

mented by a further 400–500 tons of so-called *Handkoffer*, containerized fuel, thereby enabling them to cover some 500–600 miles in all. Since it was calculated that the vehicles of an advancing army drove two miles for every mile conquered, this would have given a radius of 250–300 miles. It was intended to dump the additional supplies at points between the armoured formations and the infantry masses following in their wake, and to have the former use their *Kleinkolonnenraum* in order to replenish themselves. Mounting guard over these depôts would be the task of special detachments sent ahead by the infantry divisions.[21] By thus concentrating a great part of the available motor transport in support of the advanced spearheads OKH hoped to reach the line Dvina–Smolensk–Dnieper without a major halt. However, it was clear that this was the maximum that could be achieved.[22]

In order to go beyond the 300 mile limit the Wehrmacht would have to rely on the railways. The latter were also the only means by which it was possible to deploy the army on the Russian frontier, and work to increase their capacity was accordingly initiated in the early autumn of 1940. By April of the next year, overall capacity of the railroads crossing Poland from west to east had been increased to 420 trains in both directions.[23] This, it turned out, was in fact too much and was never fully utilized. However, the expansion of the railways was not achieved without cost. In particular, the *Eisenbahntruppe* which should have spent the winter exercising the conversion of the Russian railroads to standard gauge, were employed on other tasks and thus entered the Russian campaign without adequate training. Nor was this the only problem faced by the railway troops. Not being combat units, they figured low on the list of priorities and were finally allowed only 1,000 motor vehicles, most of which were inferior French and English material, so that a bare sixth of their formations was fully motorized and two thirds remained without any motorization at all. Dependent on the army groups to which they were allocated for their supply of fuel, the railway troops were often poorly supplied. Also, they were short of signal and communications gear, which was expected to last for the first sixty miles only.[24] Finally, their numerical strength proved hopelessly inadequate, and as early as July 1941, it became necessary to complement them with men taken from the *Reichsbahn*.[25]

To lead and supervise the entire gigantic operation, a new organization was created. This consisted of '*Aussenstellen*', or advanced detachments, of the quartermaster-general at OKH. Each army group was allocated one of these. They remained, however, independent of the field commanders and were subordinated only to Wagner himself, a solution that does not appear ideal but was probably dictated by the shortage of trained supply officers. Each *Aussenstelle* was responsible for a number of depôts, which were grouped together in 'supply districts', one per army group. Initially located on the frontier, these supply districts were to be moved forward as soon as the railways became available. To make use of captured industrial and supply installations inside enemy territory, a new branch of the Wehrmacht, the so-called technical troops, was instituted.[26] While the quartermaster-general and his various subordinates did control the magazines as well as the *Grosstransportraum*, they did not, as explained above, exercise any authority over the railways. This rested in the hands of the *Wehrmachttransportwesen*, and the number of trains to be despatched each day, as well as their destination, had to be fixed by a process of negotiation between the two authorities, each of whom would, of course, act with his own particular interests in mind.[27]

The supply-apparatus with which the Germans entered Russia was, therefore, far from impressive. At the cost of demotorizing a large part of the army and depriving it of its strategic mobility the problem of supplying the 'fast' units during the first 300 miles or so was more or less solved, but at that point logistic difficulties would certainly force a halt regardless of the situation in the field. The *Eisenbahntruppe*, whose task it was to enable the railways to take over the burden of supply at the earliest possible moment, were in some ways ill-equipped for the purpose and nowhere near numerous enough. If one railway behind each army was the normal requirement, conditions in the East were such that it was possible to construct only one line behind each army group. Even on the most optimistic estimates of consumption, stocks of some kinds were dangerously low and their supply to the front uncertain. Whether, under these conditions, even the attempt to reach Moscow – not to mention other objectives further away still – was warranted appears questionable.[28] Certainly, excessive reliance was placed on defeating the Red Army in a short time and within a reasonable distance from the frontier. Should these plans fail to be realized, supply difficulties were bound to affect the continuation of operations.

Leningrad and the Dnieper

At 0300 hours on 22 June 1941 the artillery of 144 German divisions, deployed along the Soviet border on an 800-mile front stretching from the Baltic to the northern frontier of Hungary, opened fire and thereby gave the sign for the start of the largest land-campaign of all time. Achieving complete surprise, both strategic and tactical, the German armies initially met with little opposition. Apart from isolated strongholds like Brest–Litovsk, the resistance put up by the Soviet border-guards was broken within a few hours and the way opened for the armoured and motorized divisions to race forward in order to carry out the first of the planned encirclements. Marching hard, but increasingly left behind, was the great mass of infantry and their horse-drawn transport. The often inadequately-secured spaces in between were filled only by columns of empty lorries returning from the armoured spearheads, and by parties of *Eisenbahntruppe* sent ahead to begin the task of repairing and converting the railways even before the surrounding country was well in hand. Instead of the logistic apparatus following in the wake of operations, it was supposed to precede them, a procedure probably unique in the annals of modern war, and one that is indicative of the desperate expedients which the Wehrmacht was forced to take in order to maintain its forces at all.

Until the railway lines could be used, the whole burden of supplying the gigantic operation rested almost exclusively on the *Grosstransportraum*, and here problems were encountered from the first. Though OKH had known that roads in Russia were few and bad, it was surprised by the fact that the surface of such metalled ones as existed had already begun to deteriorate on the third day of the campaign.[29] There were unmetalled roads, but during the first week of July heavy rainfall turned them into quagmires, thus confirming the fears expressed by Lossberg one year earlier. Because of the appalling roads, and also enemy action by isolated parties that had not been smashed by the advancing Panzers, losses among the lorries of the *Grosstransportraum* already reached twenty-five per cent within 19 days of the start of the campaign.[30] A week later the figure for Army Group Centre reached one third, a situation which was made worse by the fact that facilities for making major repairs were not pushed forward, but remained far to the rear in Poland and even in the Reich itself.[31]

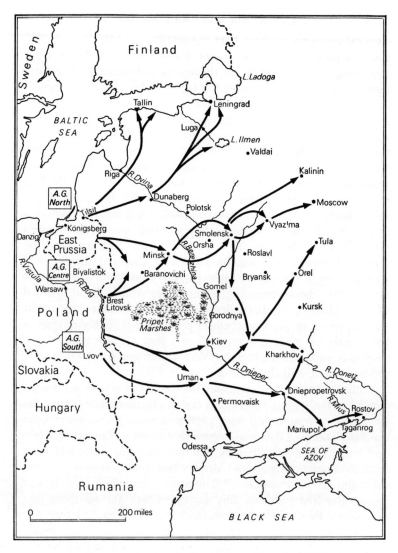

7. German operations in Russia, June–December 1941

Apart from the rapid decrease in the quantity of motor transport available, conditions in Russia also created difficulties in the operation of such vehicles as remained. There was an increased demand for fuel caused by the bad roads, rising to 330,000 tons a month (9,000 daily) instead of 250,000 as had been expected. In Russia, a normal *Verbrauchssätz* for 100 km (60 miles) lasted for 70 only.[32] The ruthless utilization of engines led to their rapid deterioration and increased the consumption of oil as against that of petrol from two to five or seven per cent.[33] Spare parts, especially tyres, were hard to obtain, because of the Reich's rapidly sinking stocks of rubber. Under such conditions the actual capacity of the *Grosstransportraum* fell far below that which had been expected, and it became difficult to supply the operation every time the front was more than sixty miles removed from the railways.[34]

In view of the different gauge of the Russian railroads, the hopes of OKH for their operation rested heavily on the utilization of captured rolling stock. Partly because the combat troops did not pay sufficient attention to the problem, however, these hopes were disappointed and it became necessary for OKH to furnish the army groups, as well as the air force supporting them, with a daily list of railways to sever inside their respective theatres of operation.[35] Nevertheless, it proved impossible to retain as large a part of the original network as had been planned. The task of conversion was heavier than expected, and the troops responsible for it were hampered by the fact that, since operations mostly followed the roads, the railways frequently had to be cleared of the enemy first.[36] This, then, was one result of the attempt to conduct a campaign with the technical means of one age and keep it supplied with those of another.

Though aided by the fact that the Russian sleepers were made of wood, not iron, the changing of the gauge was time consuming and did not solve all problems. Russian rails being lighter than German ones, per unit of length and the number of sleepers being smaller by one third, it was impossible to run heavy locomotives over converted track, which meant that use had to be made chiefly of older material. As Soviet engines were larger than German ones, water-stations were further apart, and many of these had been destroyed. Russian coal could not be used by German engines except by adding to it either German coal or petrol. Signal and communications equipment was in very short supply, having been dismantled by the retreating Soviets or, as frequently, by the advancing German troops. Nor did the railway construction battalions always carry out the repairs in a form that was acceptable from the point of view of those who would have to operate the lines. Concerned above all with restoring the greatest possible length of tracks and bridges, the *Eisenbahntruppe* frequently paid no attention to such vital matters as access to quays, workshops and engine sheds, the need to supply locomotives with coal, or even the elementary fact that one double line can carry more than two single ones. All this meant that, though General Gercke might fix the number of trains to be run each day at 48 and 24 for the two kinds of lines respectively, his figures remained largely theoretical and could never be approached in practice.[37] Finally, the conversion of Russian track created problems with the rolling-stock that was captured. Although a simple procedure could adapt wagons to standard gauge, equipment thus treated could not be used outside Russia. Locomotives were impossible to convert at all, and were therefore handed over to the Finns.

Another problem that must be considered before passing to the more detailed examination of the relationship between logistics and operations is that of the utilization of captured supplies. We have seen that Soviet fuel, both liquid and solid, was different from German fuel and could not be used without at least some preparations. Though Russian food was of course perfectly edible, the retreating Red Army did not form dumps but distributed provisions straight from the trains to the troops' vehicles, which meant that there were few magazines to capture.[38] As the Germans penetrated deeper into Russia, it became possible to make use of the resources of the conquered country, and at one point Wagner estimated that only fifty per cent of all subsistence had to be brought up from the zone of the interior. Nevertheless, provisions – even when fodder is included – formed only a small fraction of all supply requirements. However beneficial to the German economy, the utilization of Russian resources could not relieve the demands made on either the railways or the *Grosstransportraum* to any significant extent.

Of the three German army groups that entered Russia on 22 June 1941, Field Marshal von Leeb's *Heeresgruppe Nord* was the smallest and in this respect, the easiest to supply. It was also the one whose objective was nearest, the distance from its base in East Prussia to Leningrad being only about 500 miles. Compared with the rest of Russia the Baltic countries had a good road and railroad network, especially near the coast. The further north-east one went, however, the forests became denser and the roads fewer. The head of the quartermaster-general's *Aussenstelle* with this Army Group was Major Toppe. To carry out his task he had under his control (apart from the *Grosstransportraum* which was subordinated to Wagner himself) fifty lorry-columns and ten motorized supply companies with bakeries, butcheries and so on. Also at his disposal were two supply bases at Tilsit and Gumbingen, containing between them 27,803 tons of ammunition, 44,658 tons of provisions and 39,899 tons of fuel, as well as pioneering, engineering and signals equipment.[39] Leeb's forces were divided into three armies, 16th, 18th and 4th armoured (known as *Panzergruppe* 4) and comprised a total of 26 divisions, including six 'fast' ones. The section of an armoured division could be expected to consume about 300 tons of supplies a day, the rest perhaps 200. Also, Keller's I Air Fleet with some 400 planes was to be supplied.

Ordered by Hitler's directive of December 1940 to destroy the Soviet forces in the Baltic countries first of all,[40] Leeb's forces crossed the frontier at 0305 hours with *Panzergruppe* 4 sandwiched between the two infantry armies and acting as a spearhead. The pace of the advance was spectacular, with one armoured corps (von Manstein's 56th) reaching Dunaburg and seizing the crossings over the Dvina on 26 June, thus covering a distance of almost 200 miles in five days. However, the *Panzergruppe* had already outrun its service of supply. Crowded off the roads by the columns of marching infantry, the lorry-columns were brought to a standstill for days on end, with the result that, as early as 24 June, only an airlift could avert serious shortages of fuel and ammunition in both armoured corps.[41] Under these conditions the establishment of a forward supply base was an indispensable preliminary to any further advance, and this compelled the tanks to rest, immobilized, until 4 July. Even then, a fresh operation could only be mounted by concentrating the bulk of Army Group North's *Grosstransportraum* behind *Panzergruppe* 4, at the cost of bringing 16. Army to a temporary halt.[42]

Having crossed the Dvina, the two corps that made up *Panzergruppe* 4 were sent north along two divergent axes. Manstein's force moved towards Lake Ilmen with the aim of sealing off Leningrad from the east, while Reinhardt's 41st Armoured Corps

made its thrust further west towards Luga on the direct road to
Leningrad. Once again, the rate of progress was incredible. By 10
July, Reinhardt's units were standing at Luga, another 200 miles
from Dunaburg and only about 80 from Leningrad. By this time,
however, both corps were operating in terrain that was heavily
wooded and ill-suited for armour, so that the advance slowed
down, suffering from a lack of infantry. The latter had been left
far behind and was still strung out over hundreds of miles all
through the Baltic countries.

At this point, Army Group North found itself facing a dilemma.
It was clear that infantry was better suited to the type of terrain
now encountered. On the other hand, *Panzergruppe* 4 had already
announced that the supply situation would not allow it to reach
Leningrad unless *both* 16. and 18. Armies were halted in order to
enable all available transport to be concentrated behind a single
thrust.[43] Von Leeb could not bring himself to make such a radical
decision, however, and the units of *Panzergruppe* 4 therefore
found themselves almost within reach of Leningrad, but undergoing one supply crisis after the other.[44]

As the armoured spearheads plunged deeper into Russia the
Eisenbahntruppe behind them worked feverishly to restore the
railways and convert them to German gauge. By 10 July some
300 miles had thus been rebuilt, but line capacity was still so low
that, instead of the ten required, only one train per day was
reaching Dunaburg which was already hundreds of miles behind
the front.[45] For the railheads to keep up with the advance was
clearly impossible. Russian tracks with captured rolling-stock had
to be used, and it soon became clear that the transfer-points
(*Umschlagstellen*) from German to Russian trains formed the
bottleneck of the entire logistic system. Thus, heavy congestion
occurred at Eydtkau as early as 30 June.[46] Three days later, the
situation at the railway transfer point of Schaulen was said to be
'catastrophic', to the point that the shock waves went all the way
up the military hierarchy until they reached Field Marshal von
Brauchitsch, commander in chief of the army, himself.[47] Nevertheless, the station of Schaulen was again blocked on 11 July.
Instead of the regulation 3 hours, it was found that unloading the
trains took 12, 24 and even 80 hours, hopelessly congesting the
stations and making it impossible to utilize more than a fraction
of the capacity of those lines that were in use.[48] So great was the
confusion that whole trains were being 'lost', some never to re-

appear again.[49] As a result, the service of supply, though it never actually broke down, was constantly in a state of crisis. Scarcely a day passed without some unit sounding the alarm. Whereas Toppe thought he needed thirty-four trains per day (at 450 tons each) in order to meet all requirements, he actually did not get more than eighteen from the *Chef des Transportwesen* and even this number was reached only on exceptional occasions.[50]

Though *Aussenstelle Nord* repeatedly claimed that at no point had the troops suffered from an actual shortage,[51] the situation certainly was bad enough to give rise to an acrimonious debate as to where the blame lay. Distrustful of the rear services as all frontline troops are, the Army commanders – especially the outspoken Hoepner of *Panzergruppe* 4 – accused Wagner's organization of indolence and a lack of flexibility, claiming in addition that trains earmarked for his troops were 'hijacked' by 16. and 18. Armies.[52] Wagner in turn shifted the blame to Gercke who, he said, did not provide enough trains to carry all supplies. Gercke denied responsibility, pointing out the failure to unload the trains quickly. The dispute even went beyond the land army and turned into a matter for inter-service rivalry. The Luftwaffe, so Wagner's men claimed, took a larger share of the railway transport than had been agreed upon, and even had officers armed with submachine-guns riding on the trains in order to prevent outside interference.[53]

There were, however, less subjective reasons for the supply problems. Because the number of daily trains was barely sufficient to keep the front supplied, stockpiling proceeded slowly and the creation of new bases could not keep up with the advance. This placed an intolerable burden on the *Grosstransportraum*, the capacity of which had already been reduced by forty per cent because the atrocious roads prevented the use of trailers.[54] It also wasted railway capacity, as Toppe was forced to order mixed trains instead of ones that were *Rassenrein*, uniformly loaded with a single item only. In so far as the supply service admitted its inability to procure the supplies requested by the troops in less than seven or eight days, the accusation that it suffered from a certain rigidity was not perhaps unwarranted.[55] Also, it did not help matters when Toppe's organization was called upon to supply additional units from Army Group Centre.[56] However, as Wagner did not fail to remind Hoepner, the main problem stemmed from the fact that Army Group North had pushed its spearheads

forward some 400 miles in four weeks, and was operating at the
tip of a long and complex line of communications which, besides
being congested by infantry and rear echelon troops still making
their way forward, was already becoming the target of 'really
unpleasant' partisan attacks that prevented the unloading of
trains as far forward as would otherwise have been possible.[57]
Despite this, the quartermaster-general at OKH regarded the
logistic situation of Army Group North as 'the best by far' of all
the German forces in Russia, a point that was understandably
lost on von Leeb.[58]

To what extent the supply problem contributed to the German
failure to take Leningrad is difficult to say, for the planning of the
operation had been defective from the outset and its execution
marred by Hitler's nervousness and his inability to establish a
clear order of priorities.[59] Moreover, the approaches to Leningrad
were even less suitable for armour than the rest of the Baltic
countries, and on 26 July all three tank commanders – Hoepner,
Manstein and Reinhardt – unanimously recommended that it
should be withdrawn.[60] Given the fact that it was facing a
numerically superior enemy, however, it seems certain that Army
Group North's best chance for capturing Leningrad came around
the middle of July, when Reinhardt's corps had penetrated to
within 80 miles of the city. At this time, however, supply difficulties ruled out any immediate resumption of the offensive.
Having had their own way for the first fortnight of the campaign,
Hoepner's forces were meeting with stiff opposition which, in
turn, led to heavy consumption of ammunition. Demands for
extra ammunition could only be met with great difficulty, so that
stocks sank to less than fifty per cent of establishment.[61] During
the second half of July, the supply service was incapable of supporting even the most limited offensive because it was fully
occupied in moving its base forward from Dunaburg to the area
around Luga, and in this period the start of the attack was postponed no fewer than seven times.[62] So hopeless did the situation
appear that, on 2 August, Hoepner suggested the desperate
expedient of attacking the city and its two and a half million
inhabitants with a single armoured corps. The quartermaster-
general did not think he could support even this, however, and
the idea was accordingly rejected.[63] The offensive was resumed
on 8 August but by this time the defences of Leningrad were
ready.[64] Heavy rain soon turned what roads there were into

quagmires, making it impossible to meet the troops' demands for ammunition.[65] On 11 September, Hitler finally recognized the unsuitability of the terrain and ordered the withdrawal of *Panzergruppe* 4, which was to join in the final assault on Moscow, and, with the Führer's simultaneous decision that Leningrad should be tackled by the Luftwaffe, the final chance was forfeited.

To an even greater degree than Army Group North, Field Marshal von Rundstedt's Army Group South was faced with ill-defined, widely divergent objectives, which besides the conquest of the Crimea included the seizure of Ukrainian wheat, Donetz coal and Caucasian oil.[66] Consisting of forty-one divisions grouped in four armies (6, 17, 11 and *Panzergruppe* 1) Army Group South had under its command a number of Rumanian, Hungarian and Italian divisions, the maintenance of which was to prove especially difficult because these allied units were meagrely provided with motor transport and unfamiliar with German procedure.[67] Setting out from Poland south of the Pripet Marshes, Army Group South soon found itself operating in terrain that was ideal for tanks. However, there were less railways than north of the Marshes, and the heavy black soil quickly turned into a morass whenever there was a fall of rain.

Of the three army groups invading Russia, it was Rundstedt's force which found itself facing the heaviest numerical odds and, in the person of Timoshenko, the ablest Soviet commander. Progress on this front was therefore slow, and it quickly turned out that the loading of the *Grosstransportraum* with ammunition and fuel at a ratio of 1:2 was based on erroneous calculations.[68] More than any other, Army Group South was affected by the weather, and by 19 July fully half of its lorry companies were out of action. On the next day *panje* columns were organized in order to help the world's most modern army out of its supply difficulties.[69] Throughout the second half of July, there was a shortage of ammunition and bitter recriminations arose as the Army commanders accused Weinknecht's *Aussenstelle Sud* of favouritism, and each other of 'hijacking' trains.[70] Nevertheless, by the end of the month the entire western bank of the Dnieper was in German hands, but supply difficulties, coupled with the usual inability of the German infantry to keep up with its own armour, made it necessary for *Panzergruppe* 1 to waste its time reducing the encircled enemy around Uman instead of striking further across the river.[71]

Having reached this point the German advance came to a halt. Work to establish a forward supply base for an advance across the Dnieper began at once, but met with difficulties, owing to the fact that the capacity of the railways was still very low so that the lorry columns had to be driven back all the way to the Russian–Polish border. On 1 August. only four days before the new attack was due to begin, the armies comprising Rundstedt's force still had only between a sixth and a seventh of their basic load of ammunition.[72] As it was, stockpiling proceeded so slowly that it was decided to resume the offensive without waiting for the new base to be completed. This resulted in operations during August being conducted on a logistic shoestring, with shortages constantly appearing everywhere. At von Kleist's *Panzergruppe* 1 in particular, the shortage of fuel and ammunition was such that the contents of incoming trains had to be distributed in accordance with the immediate tactical situation, the details of which were, for this purpose, telephoned to the quartermaster by Zeitzler, the chief of staff.[73] By 23 August, the shortage of ammunition was becoming critical. At this moment, however, 'generous help' from 17 Army enabled the *Panzergruppe* to overcome its difficulties, at least for a time.

Acting upon Hitler's order and very much against its own will, OKH was now preparing to turn *Panzergruppe* 1 round by 90 degrees and send it north towards Kiev. In order to prepare for this move Kleist's *Grosstransportraum* was pulled out and refreshed during the last days of August, but the grievous shortage of spare parts prevented its capacity from rising above sixty per cent of establishment.[74] Even so, it does not appear that *Panzergruppe* 1 suffered from any serious problems during its operations against Budenny's South Western Theatre. The good weather facilitated the movements of wheeled transport, and the distance to the railhead at Permovaisk was not excessive. However, the performance of the two railways available to Army Group South was inadequate, the southern one becoming blocked by floods and the northern one by congestion. At *Nachschubsammelgebiet* (main supply base) South chaos reigned, and the target number of twenty-four trains per day could be met on only twelve days during the whole of September.[75] In addition, trains were arriving partially loaded.[76] The bridges over the Dnieper having been blown, any attempt to launch a serious advance to the east of it met with great difficulty, and late in September

Army Group South warned that, unless the railheads were pushed forward across the river, it would not be possible to reach the 'operational objectives' or occupy the Crimea.[77]

With the Kiev cauldron liquidated, Army Group South resumed its advance eastward on 1 October. Progress was initially very fast, and by 3 October Kleist had already completely outrun his communications which, moreover, were blocked by the usual infantry columns trudging up from the rear.[78] On 6 October the weather broke, and two days later it was slowing down Rundstedt's advance. Under 'the worst imaginable' circumstances all motor transport of Army Group South was brought to a halt (in any case, only 48 per cent of the trucks were still operational) and only *Panzergruppe* 1 on the far southernmost wing could keep going until it too was brought to a halt on 13 October. From this point onward, the situation deteriorated with extreme rapidity. On 17 October it was said to be 'catastrophic', and three days later no supplies at all were reaching the troops, who lived exclusively off the country.[79] Though no real improvement could be expected before the onset of frost, by 24 October the weather had improved sufficiently to allow the replenishment of at least one corps in preparation for the attack on Rostov.[80] Meanwhile, the troops were helping themselves by organizing *panje* columns which served surprisingly well.[81] Since there were no railway connections across the Dnieper, 6. Army improvised with sections of Russian track, whereas *Panzergruppe* 1, its spearheads now at Mariupol (subsequently renamed Zhdanov) and Taganrog, received small quantities of fuel, ammunition and spare parts by air. Far away to the rear, the railways were in an ever more disastrous state. During the whole of October, only 195 trains out of 724 scheduled actually arrived, and even this number included 112 left over from the previous month.[82]

It took some time for the true nature of the difficulties now faced by Army Group South to be realized by OKH, and when Halder returned to duty on 3 November (he had fallen off his horse and broken his collarbone) he felt that Rundstedt had become unduly pessimistic and urgently required 'a push forward'. On the next day, however, he recognized that 'all armies are stuck in the mire' from which no mere words would extricate them. There was 'no sense in putting pressure on the commanders before a solid base is established'.[83] Everyone was now waiting for the beginning of frost, but when it came on 13

November the temperature at once fell to 20 degrees (centigrade) below zero. While the roads did improve, the quantity of motor transport available fell dramatically because engines refused to start. The performance of the railways deteriorated still further, and ice floating in the Dnieper endangered the regular ferrying of supplies across the river.[84] Faced with total breakdown, *Aussenstelle Sud* began to divest itself of its functions. The armies were referred to their own resources for fuel, the *Wehrmachtsbefehlhaber* Ukraine back to the quartermaster-general at OKH.[85] That Kleist still succeeded in reaching Rostov under these conditions is remarkable; that the supply situation could be described as 'secure' during the subsequent retreat to the Mius, is even more so.[86]

In the case of Army Group South, the warning that it would not be possible to support the operation beyond the 300 mile limit was proved correct. Despite considerable difficulties, the supply service did manage to sustain the advance for as long as it was possible to employ at least part of the *Grosstransportraum* in order to relieve the railways, and even the sharp swing northward by *Panzergruppe* 1 towards Kiev did not present it with insoluble problems, since the other armies were at this time practically immobile. However, the attack over the Dnieper was started without adequate logistic preparation, long before there was the slightest prospect of pushing the railways across the river and at a time when the lines leading from Poland were the scene of great confusion which, incidentally, had nothing to do with the Russian winter. *Aussenstelle Sud* had predicted that it would not, under these conditions, be possible to reach the 'operational objectives' – which apparently meant the Donetz Basin. However, there is not the slightest hint that its view was taken into account by Hitler, OKH, or indeed Army Group South itself. Even Rundstedt, usually regarded as the most cautious of the three Army Group commanders, did not intend to suspend operations until *after* the conquest of Rostov. That this objective was itself beyond the reach of his logistic apparatus, not even he seems to have contemplated.

'Storm to the gates of Moscow'?

Of the three army groups that Hitler launched into Russia on 22 June 1941, the strongest by far was Field Marshal von Bock's *Heeresgruppe Mitte*, consisting of 49 divisions grouped in four armies (another army, von Weichs' 2nd, was kept back as an OKH reserve and brought up later), of which two (9th and 4th) were infantry and two (3rd and 2nd) armoured and motorized. The two *Panzergruppen* were subordinated to the infantry armies, a measure which was probably designed to prevent them from advancing too quickly and losing contact with the infantry in their rear. Divided into two wings, with the 'fast' formations on the extreme left and right, it was the task of Army Group Centre to defeat the enemy on its front by means of three *Kessel-schlachten*, with the last pair of pincers closing in at Smolensk. At this point, mobile operations were to come to a halt.[87]

Having overcome the resistance put up by the Soviet frontier troops, Bock's advance gained momentum during the morning of 22 June and the two *Panzergruppen*, particularly Guderian's 2nd on the right wing, found themselves advancing rapidly into the depths of Russia. Though the country was less suitable for armour than the Ukraine, it was nevertheless much easier than that faced by Hoepner farther north. However, roads were few and marching discipline poor. Huge masses of infantry blocked the bridges over the Bug, and by the evening of 25 June the *Gross-transportraum* earmarked for Guderian's support had still not succeeded in crossing the river. As early as 23 June, Guderian was therefore obliged to request the supply of fuel by air.[88] Similar problems arose in the area of 9. Army, whose infantry and the supply columns of Hoth's *Panzergruppe* 3 were struggling for priority on the roads. Fuel consumption by both armoured groups was very high, but could be met because that of ammunition was correspondingly low, and because *Panzergruppe* 2 made a timely discovery of a large Russian reservoir near Baranovichi.[89] Scarcely any subsistence at all was making its way forward, but the troops found it possible to live off the country. On 26 June, Guderian and Hoth closed the first of their pockets at Minsk, while the infantry in their rear was at the same time building a smaller pocket at Biyalistok. On 16 July the armoured groups met again, this time at Smolensk. No very great difficulties of supply appear to have been encountered during these operations, although the distances covered were such that, already on the tenth day of the campaign, some tanks were lost because of a shortage of spare parts.[90] Moreover, the encirclement at this time was a very sketchy affair; it would be weeks before the infantry could catch up, and in the meantime the armoured formations were compelled to stay almost immobile, beating off counter-attacks and fretting over their inability to push forward. Since the fighting now bore a defensive character, the consumption of fuel fell sharply while that of ammunition rose to extraordinary heights, giving rise to more than one critical moment.[91]

Meanwhile, there were problems on the railways which, at such distances from the border, could alone guarantee the service of supply for any length of time. The road-bound German method of fighting had left much of the length of the railways untouched, and there were far too few security-troops to deal with the situation. For this reason, and also because transfer points from German to Russian rolling-stocks formed bottlenecks, performance fell so far below expectations that 9. Army complained that it was only receiving a third of the daily contingent of trains to which it was entitled.[92] Instead of improving, the situation grew worse. After 8 July, the railway only carried supplies for *Panzergruppe* 3 and 9. Army had to use its *Grosstransportraum* even though the distance from base now exceeded 250 miles and the roads in between were atrocious.[93] As always, it took some time for OKH to grasp what was going on. On 13 July Wagner optimistically reported that he could supply the armoured groups for an advance to Moscow, but next day he modified his estimate, in so far as he now admitted that they could go no farther than Smolensk, and that the infantry would have to halt even further to the west, on the Dnieper.[94]

From the middle of July, the supply-situation of Army Group Centre was developing signs of schizophrenia. On one hand Wagner and Halder were aware of some 'strain', but nevertheless confident of their ability to build up a new supply basis on the Dnieper, from which further operations were to be launched around the end of the month. They appeared not to hear the loud cries for help from the armies. The consumption of ammunition throughout this period was very high, and could be met only – if at all – by means of a drastic curtailment in the supply of fuel and subsistence.[95] 9. Army was fighting around Smolensk, but its nearest railhead was still at Polotsk – and this at a time when a basic load of fuel lasted for only twenty-five to thirty miles instead of the regulation sixty-five.[96] Around the middle of August, both 9. and 2. Armies were living from hand to mouth, with stocks of ammunition still falling instead of rising in preparation for a new offensive. Moreover, the supply of POL (petrol, oil and lubricants) was quite insufficient, and did not take into account the worn state of the engines.[97] The continued resistance of the Russian troops trapped inside the Smolensk pocket delayed the refreshment of the armoured formations and this finally required almost a month, instead of the three or four days Guderian thought would be needed. Even then it remained incomplete, for Hitler refused to make available new tank engines at the expense of the program of vehicle production.[98] Meanwhile, even though the Smolensk pocket had finally been liquidated, Army Group Centre was still engaged in heavy fighting. Throughout August it had to face enemy counter-attacks from the east, causing heavy expenditure of ammunition that could only be met by cutting back on subsistence. Stockpiling for a new offensive was impossible.[99]

As Bock's forces were coming to a halt, OKH and Hitler had very different ideas about the continuation of operations. The former talked of decisively beating the Red Army by advancing to Moscow, the one objective from which the Russians would find it impossible to retreat; the latter was more interested in Ukrainian wheat, Donetz coal and steel, Caucasus oil, and the capture of the Crimea ('that aircraft carrier against the Rumanian oilfields'). Against the proposals of OKH for the continuation of the offensive in the direction of Moscow, Hitler argued that the Russians simply ignored threats to their rear and went on fighting – a theory that was borne out by the fact that, in every one of the previous battles, the Red Army had continued to resist even when surrounded and often succeeded in getting out large bodies of troops from the thinly-held pockets. Given this relative immunity to encirclement, Hitler argued that the previous German operations had been too ambitious. The way to destroy Russia's 'living force' was to proceed slowly and methodically, driving the Russians into successive small pockets and eliminating them one after the other. For a start, he proposed to liquidate what the Germans insisted on calling 5 Soviet Army near Kiev, though in fact the forces involved numbered at least four Armies and parts of two more.[100] The greatest advantage of this operation was that it could be carried out by Army Group South in conjunction with *Panzergruppe* 2, which had now finally been extricated from the battles round the Smolensk pocket and was at least partly refreshed. The rest of Army Group Centre would not be involved, which in view of its logistic situation was just as well.

To judge from the records of *Panzergruppe* 2, Guderian's forces did not – thanks mainly to the German railway reaching Gomel late in August – suffer from any very great supply difficulties during the southward thrust to Kiev. On the other hand, the need to support this operation did have some negative effect on the replenishment of 2. Army, which was leading a precarious hand to mouth existence at the end of a Russian railway line from Gomel to Gorodnya. The capacity of this line was initially small. It had just started showing signs of improvement when, on 12 September, floods blocked the roads from the railhead to the corps and forced its operation to be suspended. Under these circumstances, it was only on 15 September that 2. Army could again describe its supply situation as 'secure' (*gesichert*).[101] Stockpiling got slowly under way again, and seems to have been more or less complete by the end of the month. However, the logistic situation was such as to preclude an attack by the Army before the beginning of October.

Further to the north, the situation of the other units making up Army Group Centre was not dissimilar. To meet current consumption and build up stocks for the attack on Moscow, Bock estimated that he needed 30 trains daily. Gercke only promised 24, however, and the average number for the first half of August did not in fact exceed 18. After the conversion to German gauge of the Orsha–Smolensk railroad on 16 August the situation showed some improvement, but the target number of 30 trains per day was never reached.[102] Though this was to be the final and decisive act of the Russian campaign, OKH did not concentrate all its resources behind it. On 15 August a nervous Hitler ordered Hoth's *Panzergruppe* 3 to send out a corps in order to help Army Group North, a move which not only led it into very difficult country where the 'use of tanks was very stupid', but also created supply difficulties because it involved a change of direction by almost 180 degrees.[103] Nevertheless, Brauchitsch seems to have agreed with Hitler on this occasion. What is more, he ordered 5,000 tons *Grosstransportraum* to be diverted from Army Group Centre to Army Group South, where stockpiling was meeting with great difficulties.[104] In view of this almost unbelievable dispersion of resources, 9. Army on 14 September flatly declared that its transport 'was insufficient to support the coming operations'.[105] The commander of 4. Army, von Kluge, took a personal interest in the state of supply when he wrote:[106]

> The supply situation of the Army may, on the whole, be regarded as secure. . .with the growing distances, the Army is almost completely dependent on the railways. At the moment, the latter meet current consumption only. The transport situation did not so far allow the establishment of depôts sufficiently large to enable the troops to receive what they need in accordance with the tactical situation. The Army lives from hand to mouth, especially as regards the fuel situation.

After being interrupted for eight days because of floods,[107] stock-building for Army Group Centre was resumed on 21 September, and by the end of the month was more or less complete. However, this could only be achieved by cutting back on subsistence, so that the troops were forced to live off the country. Other shortages included motor oil (the bottleneck of the entire transportation system), vehicles, engines, spare parts for tanks (which, on Hitler's orders, were not being manufactured at all) and tyres, which only arrived at the rate of one a month for every sixteen vehicles.[108] Fuel was so short that the shortage threatened to bring the operation to a halt in November. This was due partly to its non-availability at home, and partly to the impossibility of supplying six armies (including three armoured ones – Hoepner's *Panzergruppe* 4 had now joined the Army Group Centre) with some seventy divisions between them, at a distance of 400 miles from their bases.[109]

After getting off to a late start on 2 October, the German attack on Moscow was at first as successful as during the previous offensives. Operating in the usual manner, Hoth and Hoepner each formed a prong of a pincer-movement that snapped shut at Vyaz'ma on 8 October, trapping some 650,000 Russians. At the same time, Guderian in the far south was making good progress in his efforts to outflank the defences of Moscow from the right. There were problems of supply from 4 October when *Panzergruppe* 4 complained that it had begun the operation with only 50 per cent of its motor transport still serviceable.[110] Four days later 4. Army protested against the small number of fuel-trains arriving from the rear.[111] Between 9 and 11 October, the weather broke; rain turned the countryside into a morass, and the few roads that were available quickly broke up under the weight of traffic. From this time for about three weeks all armies were stuck in the mire, unable to move either forward or backward, and the troops reduced to living on whatever could still be taken away from the country. In the autumn mud, Hitler's soldiers floundered. For the success of its offensive, the world's most modern army now depended on small parties of infantrymen, unsupported by heavy weapons and accompanied only by *panje* carts.

Although Army Group Centre remained stuck in its place until frost set in around 7 November, this does not mean that the situation was everywhere uniformly bad, or that no local improvements occurred, enabling at least some supplies to go through. Moreover, close scrutiny of the quartermasters' diaries reveals what divisional histories tend to conceal, namely that the difficulties were due as much to the poor performance of the railways as to the ubiquitous mud. Since the crisis in railway transportation (especially of fuel) began well before the onset of frost, this goes some way to correct the impression that the German failure to take Moscow stemmed solely from the lateness of the season.[112] Thus, at Guderian's *Panzergruppe* 2, the state of the roads led to grave supply problems from 11 October onward. At the same time, however, the number of fuel-trains reaching Orel fell very sharply, thus making it impossible to resume the offensive even after frost had hardened the roads and the tactical situation had once more become 'favourable'.[113] At Strauss' 9. Army, only four fuel-trains arrived in the twenty days from 23 October to 13 November, in spite of the fact that mild frost (5 degrees centigrade below zero) did not set in till 11 November, and remained at this level for several days more.[114] South of the main motorway from Smolensk to Moscow, *Panzergruppe* 4 was still advancing slowly as late as 25 October, driving 'a weak opponent' before it and imploring OKH to make 'ruthless use' of the railways in order to provide it with fuel.[115] At 2. Army, the situation first became serious on 21 October. Its artery of supply, the road from Roslavl' to Bryansk, had deteriorated and out of the 3 daily trains that the Army was demanding at either Orel or Bryansk only one arrived. Weichs thereupon warned that his supply situation would become desperate unless trains arrived, an admonishment that was repeated day after day until the end of the month.[116] Alone among the Army commanders, von Kluge repeatedly asserted that his stocks were large enough, but the difficulty lay in bringing them to the troops, for which purpose the railway to Vyaz'ma was coming into operation from 23 October onward. There followed a few critical days, but on 28 October the supply situation was once again said to be 'secure', and there is evidence for this in the reserves then available to the troops.[117] Frost in this sector seems to have started earlier than anywhere else, thus improving conditions on the roads and enabling the Army to assert repeatedly, between 6 and 8 November, that the supply situation was 'secure, also in view of the coming operations'.[118] By 13 November Kluge had lost some of his optimism, claiming that Eckstein's *Aussenstelle Mitte* was favouring other armies at his expense.[119] It seems certain, therefore, that the mud was only one factor that brought the Wehrmacht to a halt. No less important were the railways, which had already experienced such tremendous difficulties in building up a base at Smolensk and which were simply unable to cope with the increased demands of a fresh offensive.

After the middle of November, the relative importance of these facts became more obvious. Frost had now set in everywhere, making the roads passable once more, though hewing the vehicles out of the mud in which they were stuck up to their axles was a difficult process in which many of them were irreparably damaged. Ignition systems, oil and radiators also gave trouble, though in theory at least all army groups had been supplied exclusively with freeze-proof POL from 11 October onward.[120] However, it was on the railways that the cold had its worst effects. German locomotives did not have their water-pipes built inside the boilers so that seventy to eighty per cent of them froze and burst.[121] The transportation crisis that followed was far greater than anything that had gone before. Between 12 November and 2 December hardly any trains at all arrived for 2. Army, leading to grave shortages of every description that appear to have had little to do with the state of the roads.[122] From 9 until 23 November only one fuel train for 9. Army arrived, and the contents of this one could not be distributed because the lorries' tanks were themselves empty. Nevertheless, during this entire period, the *Grosstransportraum* were consistently more efficient than the railways, bringing up considerable quantities of supplies and enabling the Army somehow to hold out.[123] With *Panzergruppe* 4, supplies from the rear – especially fuel – simply did not arrive after 17 November.[124] As was the case in the previous month, 4. Army seems to have been the exception. Its *Grosstransportraum* had fallen to a bare eighth of its original strength, but supplies were reaching it by rail in more or less adequate quantities.[125]

Meanwhile, far away in East Prussia, Hitler and OKH were considering the situation. On the evening of 11 November a meeting was held, in which the Führer not only confirmed his intention to capture Moscow, but also set objectives far beyond the city. Visiting Army Group Centre two days later, Halder faced vehement protests by Eckstein. Bock, however, did not support his supply officer, insisting that to make one final effort was preferable to spending the Russian winter in the open. Even though it was clear to him that the operation could not be adequately prepared, Halder reluctantly allowed himself to be persuaded by Bock, saying that he would not rein in Army Group Centre if it wanted to attempt the attack, for an element of luck belonged to warfare.[126] Thus, the final attack was authorized, only to fail, in the first place, because of the state of the railways.

Before we leave the German forces in their sorry state in front of Moscow, it is necessary to say a word about the much-discussed problem of winter equipment. That Hitler forbade his commanders even to mention the subject may be true, though this did not prevent Halder from taking a preliminary look at the problem on 25 July.[127] Anyone who has studied the documents cannot fail to be impressed by the hundreds upon hundreds of orders, directives and circulars concerning winter supplies that began to emanate from OKH from early August onward, covering every detail, from the reconnoitring of suitable shelters to the provision of freeze-proof POL, from winter clothing to veterinary care for horses.[128] To what extent these documents represent a concrete reality is very difficult to say, but there is no reason to suppose that OKH was engaging in mere mental gymnastics. Moreover, we have the evidence of Wagner and his subordinates that winter equipment was available in 'sufficient' quantities, but could not be brought up owing to the critical railway-situation.[129] It is certain that the railroads, hopelessly inadequate to prepare the offensive on Moscow and to sustain it after it had started, were in no state to tackle the additional job of bringing up winter equipment. Therefore, the question as to whether or not such equipment was in fact available is perhaps of secondary relevance.

Conclusions

The German invasion of the Soviet Union was the largest single military operation of all time, and the logistic problems involved of an order of magnitude that staggers the imagination. The means with which the Wehrmacht tried to tackle these problems were extremely modest. If it came so close to its goal, this was due less to the excellence of the preparations than to the determination of troops and commanders to give their all, to bear the most appalling hardships and to make do with whatever means were given to, or found by, them.

For the Russian campaign, the Wehrmacht never had sufficient means available, and this was even more true of raw materials, reserve stocks and means of transportation than it was of combat forces. It has been estimated that, in order to reach Moscow – not to mention the line, still further away, from Archangelsk to the Volga – by means of motor transport alone, at least ten times the number of vehicles actually available would have been needed.[180] At the same time, the *Eisenbahntruppe*, on whose shoulders the main burden of the logistic apparatus ultimately rested, were by no means sufficiently numerous, as well as being in some ways ill-equipped and ill-trained. That the railways were far from being a sufficiently flexible instrument to be able to give backing to a *Blitzkrieg*, the events of 1914, or even those of 1870, had amply demonstrated. Yet even by disregarding the entire railway network and concentrating all its resources on motor vehicles, the Wehrmacht could not have even approached a degree of motorization sufficient to enable it to carry on war against Russia by means of motor transport alone.

Thanks to the fact that it concentrated the bulk of its motor vehicles behind the four *Panzergruppen*, and also because the infantry did not have too much fighting to do during the early stages of the campaign, the Wehrmacht succeeded in driving its spearheads to Luga in the north, the Dnieper in the south and Smolensk in the centre. At these points operations came to a halt, as had indeed been expected to happen even before the campaign was launched. At Army Group North, the building of a new base took so long as to nullify any prospect of taking Leningrad. At Army Group South, it was so difficult as to make it necessary to start the next offensive without any proper base at all, with the result that operations east of the Dnieper always hung by a thread and ultimately came to a halt short of the operational objectives. At Army Group Centre, the construction of a forward base took the best part of two months and even then some crucially important items, including above all spare parts, tyres and engine oil, remained in short supply. As to the 'supply' of tyres, this was so small as to merit one adjective only – ridiculous.

There is no doubt that the logistic situation would not have allowed an advance by Army Group Centre on Moscow at the end of August. At the very best, a force of between 14 and 17 armoured, motorized and infantry divisions might have been so employed,[131] and whether this would have been enough, even in September 1941, to break through the city's defences is very much open to question. It is arguable, moreover, that since the approaches to Moscow were less suited to mobile warfare than was the Ukraine, not even *Panzergruppe* 2 could have been supplied. By preventing Guderian from being sent to Kiev, OKH would have spared his tanks much wear and tear and perhaps made it possible to speed up the replenishment of 2 Army. The main forces of Army Group Centre would not have been affected, however, for their supplies came through another railway-line. The performance of this line was such that, even as late as 26 September, fuel stocks of Army Group Centre were actually falling. The delay imposed by Hitler's decision to give the Ukraine priority over Moscow was therefore far shorter than the usual estimate of six weeks. The postponement, if there was one, can hardly have amounted to more than a week or two, at the very most.

The difficulties experienced in building up a base for the attack on Moscow also rule out another suggestion that is sometimes made, namely that Hitler, instead of dissipating his forces in simultaneous offensives along three divergent axes, ought to have concentrated them for a single attack against Moscow. The logistic situation ruled out such a solution, however, for the few roads and railroads available would not have allowed such a force to be supplied. Even as it was, the concentration of seventy divisions for the attack early in October gave rise to very great difficulties, especially with the railways and the supply of fuel. It would have been utterly impossible to construct an adequate forward base for a force twice that size.

Among the factors that prevented the Germans from entering Moscow, general mud is usually considered the most important. It is true that the weather delayed the Germans by two to three weeks, but it must be remembered that a crisis in railway transportation had been developing well before the onset of the *rasputitsa*, the season of slush. During October, railway performance was hopelessly inadequate, and supplies of fuel almost nonexistent, owing to shortages in the Reich. Had it not been for this breakdown of railway transport, it is probable that Bock could have resumed his attack by up to a week earlier than was actually the case. In the event, frost when it came hit the railways more than it did motor transport, and whereas the latter was still giving valuable – if limited – service in November, the former was reduced to almost negligible proportions by a short-age of locomotives.

In view of the undoubted importance of mud as a factor in the German defeat, it has been suggested that the Wehrmacht was wrong in basing its logistic system on wheels instead of tracks.[132] It is true that only tracklaying vehicles could have negotiated the approaches to Moscow in October. However, to suggest that all the 3,000-odd vehicles of the armoured divisions ought to have been of this kind is to misunderstand completely the working of the German war machine during this period. Even if they had been capable of producing so many tracked vehicles, which of course they were not, the Germans would have been hopelessly unable to provide them with fuel and spare parts, both of which were in desperately short supply. Indeed, such are the demands of tracked vehicles in these two respects that, even in today's world, which is capable of feats of production far in excess of the Wehrmacht's wildest dreams, there is not a single army anywhere in the world that carries all, or even most, of its supplies on such vehicles.

Since the Wehrmacht that set out to conquer Russia in 1941 was a poor army with strictly limited resources, success – from the logistic point of view – depended above all on a correct balance between railways, wheels and tracks. A detailed scrutiny of the logistic system, as well as the plain fact that the German victories of 1941 were among the greatest of all time, seem to in-dicate that, by and large, this balance was achieved and that the solution which was actually adopted was probably the best possible. Had politico-military-economic considerations allowed Germany to attempt the conquest of Russia in a slow and method-ical manner, more reliance could have been placed on the rail-ways. Had a very much stronger motor industry existed, wheels and tracks could have played a larger role. However, during the whole of World War II there was only one belligerent that could even begin to create a fully motorized army – the United States.

This should not be taken to mean that, within the limits of the resources available, German planning and organization were always ideal. This was far from being the case, as is shown by the following examples. The division of the transportation system between two authorities, the *Chef des Transportwesen* and the quartermaster-general, with only the chief of the general staff to coordinate their functions, was ill-conceived and led to endless friction. The structure of the quartermaster-general's organization was also unsatisfactory, for it deprived the Army Group commanders of their own supply-apparatus and left them sandwiched uneasily between OKH on one hand and the Army quartermasters on the other, the latter receiving their orders not from the commanders but from Wagner's *Aussenstelle*.

In planning the campaign, too much reliance was placed on the capture and utilization of Russian rolling-stock. When locomotives and wagons failed to materialize in the desired quantities, it became necessary to convert the lines to German gauge, an operation which was not technically very difficult but which required more *Eisenbahntruppe* than the Germans had available. The result was that both German and Russian lines had to be used, the transfer-points becoming bottlenecks that were constantly being pushed forward but never eliminated. Coordination between the headquarters responsible for reconstructing the railways and those who operated them was faulty, with the former ignoring the requirements of the latter.

Planning and control of the railway transports into Russia were far from perfect. The governor of Poland, Frank, was uncooperative, and it was not until November 1941 that the army finally succeeded in having its demand for absolute priority for military trains accepted.[133] There was not enough personnel to unload the trains, and prisoners of war had to be used for this task.[134] Traffic control was decidedly lax, so that some trains were 'hijacked' and others completely lost. Communication with local personnel was difficult, because of the language problem. Between the time a railway was completed and its becoming operational, long and unwarranted delays often occurred.[135] The system lacked flexibility, especially regarding the cancellation of trains.[136] The forces allocated to guarding the lines were totally inadequate.

As the number of field police was far too small, road-traffic control was at all times poor, and sometimes led to considerable friction, especially at the beginning of the campaign – when the *Handkoffer* earmarked to replenish the 'fast' formations remained in the same place for days on end – and during the battle of Moscow, when competition for the few available roads reached catastrophic dimensions.[137] The motor transport companies were badly organized, not enough of them being concentrated in the hands of OKH.[138] The troops were continually trying to overcome the shortages of material by bypassing the regular supply channels, an evil that could never be entirely eradicated. On the other hand, the same troops displayed a marked reluctance to cooperate with the service of supply, the result being that their organic vehicles were not sufficiently utilized and, even though they were exposed to enemy action, actually suffered fewer losses than did the overworked *Grosstransportraum*. The quartermasters at army level were unable to exercise proper control over the division of booty, the result being that some units saturated themselves with captured vehicles, whereas others, especially the rear services, suffered from grave shortages. The loading of the motor transport had initially been based on erroneous calculations, there being too much ammunition and too little fuel. The result was a shortage of the latter, whereas the former had to be left lying in the field because the trucks that were supposed to carry it could not get their tanks replenished.[139] That the utilization of the lorry companies was not always perfect is shown by the incredible diversion of 5,000 tons of precious *Grosstransportraum* from Bock to Rundstedt, at the very moment when the former was about to begin his decisive offensive against Moscow.

Back in Germany, calculations as to the loss of vehicles to be expected were hopelessly optimistic. It was intended to conduct the operation without bringing up any replacements as all.[140] The relative priorities allowed to the production of self-propelled vehicles – including tanks – on one hand, and their spare parts on the other, were ill-advised. This was the result of Hitler's fascination with numbers, and his insistence that establishing new formations was preferable to the bringing up to establishment of existing ones.[141] Spare parts were therefore in short supply, and the arrangement by which they were only issued to the troops *en lieu* for used ones led to constant friction with the supply service.[142] The organization of the repair-service was also faulty, since a large part of it remained in the Reich on the assumption that the campaign would be over before a major overhaul became necessary.[143]

While listing all these shortcomings, it is essential not to lose a sense of perspective. Logistics form but one part of the art of war, and war itself is but one of the many forms that political relationships between human societies may assume. Whether Germany was strong enough, in 1941, to defeat Russia while waging a war on two fronts may well be doubted.[144] However, it is difficult to see what other way was open to Hitler, after his failure either to reach a political settlement with Britain or to render her harmless by military means.[145] Risky the war against Russia may have been, but there can be little doubt that it was essential for the Third Reich's survival, even if one does not believe that a Soviet attack was imminent. This war was lost on grounds other than logistic, including a doubtful strategy, a rickety structure of command and an unwarranted dispersion of scarce resources. While recognizing the magnitude of the achievement – among other things, logistic – that brought the Wehrmacht almost within sight of the Kremlin, the above-listed factors certainly played an important role in its failure, and for this it is OKH, not Hitler, who must be held responsible. In logistics, as in everything else lying between minor tactics and strategy, the Führer had no interest whatsoever. Apart from one or two points, any errors that were committed in these fields – which is said to comprise nine-tenths of the business of war – must be laid squarely at the door of Halder and the General Staff. Even the most important decision Hitler made during the campaign of 1941, namely the sending of Guderian into the Ukraine instead of towards Moscow, was justified on logistic grounds and certainly had little to do with the postponement of the drive on the Russian capital. In war, it is often the small things that matter; and in many of these the Wehrmacht had been weighed, counted, and found wanting.

6

Sirte to Alamein

Desert complications

The question whether an Axis advance into the Middle East
could have won the war for Hitler is still one of the most contro-
versial in the history of World War II. Whereas earlier writers
claimed that by supporting Rommel in a drive from Libya
through Egypt, Palestine, Syria and Iraq to the Persian Gulf,
Hitler could have gone far towards winning the war against
Britain,[1] more recent scholars have questioned this view and
assert that, in the final analysis, the Führer was right in refusing
to regard the Mediterranean as anything but a secondary
theatre.[2] Whatever their differences, both schools agree that the
problem was essentially one of Hitler's volition. That is, the
question was not whether he could have sent more forces to the
Mediterranean but whether he should have done so. This, however, is by no means self-evident. While Rommel in his memoirs
has cast the blame for failing to solve his supply problem very
widely, the man responsible for coordinating those supplies – the
German military attaché in Rome – has written an article claiming that the problem was insoluble in the first place.[3] Since,
however, both accounts are cursory, and their authors hardly dis-
interested, the question remains whether the aforementioned
objectives were within reach of the Axis forces.

To begin with, the problem itself needs to be clearly defined.
First, although Hitler and his staff did in fact have plans for the
occupation of Gibraltar and even for seizing French North-West
Africa with the adjacent islands,[4] we shall assume that it was in
the eastern Mediterranean, if at all, that the war against Britain
could have been won. Second, it is assumed that any Axis advance
into the Middle East would have been limited to the south, i.e.
Libya and Egypt, because an attempt to go through Turkey

would have met with Soviet resistance and developed into a German–Soviet war.[5] These two assumptions enable us to ignore most of the political difficulties involved in Germany's cooperation with Italy, France, Spain and Turkey, and to focus on the question whether a German–Italian advance from Libya into Egypt and the Middle East was militarily feasible.

Rommel's first offensive

In the annals of military history the campaigns waged in the Western Desert are often said to occupy a unique place, and nowhere is this more true than in the field of supply. By and large, the story of logistics is concerned with the gradual emancipation of armies from the need to depend on local supplies, though we have had ample occasion to see that this development was by no means a straight and simple one. Even a Guderian in Russia and a Patton commanding a motorized army in France were able to make direct use of some local resources at least, and behind both there came vast administrative machineries whose purpose it was to organize the zone of communications and exploit it in the interest of the war-effort as a whole. Operating in the desert, neither the British nor their German opponents had the slightest hope of finding anything useful but camel dung, and while the former did at least possess a base of some considerable size in Egypt, the latter were entirely dependent on sea-transport even for their most elementary requirements. For just over two years, every single ton that was consumed by Rommel's troops had to be laboriously crated in Italy, then shipped across the Mediterranean. Ammunition, petrol, food, everything was brought up in this way, and such were conditions in the desert that even water often had to be transported over hundreds of miles.

Added to this problem were the enormous distances that were out of all proportion to anything the Wehrmacht had been asked to deal with in Europe. From Brest-Litovsk, on the German–Soviet demarcation line in Poland, to Moscow it was only some 600 miles. This was approximately equal to the distance from Tripoli to Benghazi, but only half that from Tripoli to Alexandria. Apart from odd bits of 95cm. track,[6] these vast empty spaces had to be entirely covered by road, and even of these there was only one – the Via Balbia stretching endlessly along the coast, sometimes liable to be interrupted by floods and always a convenient target for aircraft roaming overhead. Apart from this there were only desert tracks, which, though they had perforce to be used, subjected the vehicles traversing them to greatly increased wear and tear.

Wholly unaccustomed to desert warfare as they were, the Germans faced some further problems that reflected their lack of experience. Their diet, for example, was unsuited to the African heat, its fat-content being too high. Partly as a result of this, it was considered impossible to have a soldier stationed in Libya for more than two years without his health being permanently affected. German engines, especially those of motor cycles, tended to overheat and stall. Tank engines also suffered, their life being reduced from 1400–1600 miles to only 300–900.[7] Some of the effects of the heat and poor roads might have been prevented by strict maintenance, but this was not a field to which Rommel paid the greatest attention. In any case, maintenance was especially difficult because German and Italian equipment was not standardized.[8] Looking back some years later, Rommel wrote that a general should take personal care of his supplies in order to 'force the supply staffs to develop their initiative'.[9] In practice, however, this only meant that in the balancing of operational prospects against logistical possibilities the latter were frequently ignored.

The sending of Wehrmacht units to North Africa was first given serious consideration in the early days of October 1940.[10] A German staff officer, General Ritter von Thoma, was detached to the Italian army advancing into Egypt in order to study conditions on the spot. On 23 October he reported that only motorized forces were of any use in the desert. To ensure success, 'nothing less than four armoured divisions would suffice', and this was also 'the maximum that could be effectively maintained with supplies in an advance across the desert to the Nile Valley'. This small force would have to be of the highest quality, he said, implying the replacement of the Italian troops in Libya by German ones, a step to which, as Hitler well knew, Mussolini would never give his consent.[11] No decision was made, and with that the matter ended for the time being.

In January 1941 it arose again. Instead of penetrating further into Egypt, the Italians were being thrown out of Cyrenaica by Wavell's Nile Army. Hitler, although he felt that even the loss of the whole of North Africa was 'militarily tolerable' to the Axis,[12]

sufficiently feared the political repercussions of such a development on Mussolini's position to want to send a *Sperrverband* to
help stem the British advance. Since, as his staff hastened to point
out, such a unit would ultimately have to be taken away from the
forces earmarked for the projected invasion of Russia,[13] the
Führer resolved to keep it as small as possible.

In the event, the maintenance of even this small force proved
problematical from the beginning. Some personnel and a limited
quantity of supplies could be flown in. Early in 1942, no less than
260 aircraft – including a number of giant ten-engined hydro-
planes – were thus engaged. However, the bulk of the *materiel*
would have to go by sea. With Naples, Bari, Brindisi and Taranto
all available as ports of embarkation, little difficulty was to be
expected at the Italian end of the crossing, though the structure
of the Italian railway network was such that most transports were
confined to the first of these ports. However, following their
retreat from Cyrenaica, the Italians, in February 1941, were
reduced to a single port for unloading supplies. This was Tripoli,
the largest Libyan harbour by far, capable of handling – under
ideal conditions – five cargo ships or four troop transports
simultaneously. Its capacity, as long as no unforeseen explosions
wrecked the quays, and the largely local labour-force was not
driven off by air raids, amounted to approximately 45,000 tons
per month.

At Tripoli, however, the problem of maintaining an army in
North Africa was only beginning. On operational grounds, Hitler
wisely made his agreement to help Mussolini in Africa conditional
on the Italians holding not just Tripoli and its immediate surroundings, as they had originally intended to do, but a considerable area that would enable his forces to manoeuvre and afford
some protection against air attacks.[14] This decision, together with
Churchill's withdrawal of part of Wavell's force for employment
in Greece, led to the front being stabilized at Sirte, 300 miles east
of Tripoli. Since there was no adequate railway running eastward
from Tripoli this meant that, even under the most favourable
circumstances, the German force would have to operate at a
distance from its base half again as large as that normally considered the limit for the effective supply of an army by motor
transport.[15] Instigated by his generals, Mussolini ventured to
draw Rintelen's attention to this fact; the Germans, however,
chose to override him, thus creating for themselves that clash

between operational and logistic considerations that was to bedevil their presence in Africa to the end.

A motorized force of one division, such as the Germans originally sent to Libya, required 350 tons of supplies a day, including
water. To transport this quantity over 300 miles of desert, the
Army High Command calculated that, apart from the troops'
organic vehicles and excluding any reserves, thirty-nine columns
each consisting of thirty two-ton trucks would be needed.[16] This,
however, was only the beginning. Rommel had scarcely arrived
in Tripoli when he started clamouring for reinforcements and
Hitler, overriding Halder's objections, decided to send him the
15th armoured division. This raised the motor-transport capacity
needed to maintain the Deutsches Afrika Korps (DAK) to 6,000
tons; since this was proportionally ten times as much as the
amount allocated to the armies preparing to .invade Russia, the
announcement was met by howls of protest from the OKH
quartermaster-general who feared lest Rommel's insatiable requirements would seriously compromise operation Barbarossa.[17]
Moreover, should Rommel receive still more reinforcements – or
should he go beyond the 300 mile limit – a shortage of vehicles
was bound to ensue. Coastal shipping, it was found, could not
significantly alleviate the problem; while granting Rommel his
trucks, therefore, Hitler coupled them with an explicit order
forbidding him from taking any large-scale offensive action that
would raise his requirements still further.[18]

Even without an offensive, however, Rommel's demand for a
second division had already jeopardized his supplies. Together
with the Italians, the Axis force in Libya now totalled seven
divisions which, when air force and naval units were added,
required 70,000 tons per month.[19] This was more than Tripoli
could handle effectively, so that a crisis was bound to develop
unless the French agreed to allow 20,000 tons of supplies a month
to pass through their port of Bizerta.[20] Although Rommel was
usually at loggerheads with his nominal Italian superiors, they
were in agreement this time, for Mussolini had long been looking
for just such an opportunity to penetrate Tunisia. Hence
Rommel's request was enthusiastically seconded.[21]

Negotiations with Vichy were accordingly initiated. Firstly, the
premier, Admiral Darlan, was asked to sell the Germans French
lorries stationed in Africa, to which he immediately agreed.[22]
Encouraged by this success, Hitler next summoned Darlan for a

tête a tête on 11 May, in the course of which he told him that the unloading facilities of Tripoli harbour were 'being used to capacity' and asked for permission to use Bizerta. Darlan acceded to the request, and on 27–8 May a German–French protocol was signed in Paris, granting the Germans rights of transit through Bizerta. It also provided for French ships to be chartered by the Axis, and mentioned Toulon as a possible alternative port of embarkation in case Naples became choked.[23] At this point, however, Vichy was alarmed by the British invasion of Syria. For reasons of their own, the Germans also came to regret the agreement,[24] and by the end of the summer not a single Axis load had passed through Bizerta.

Meanwhile Rommel, defying Hitler's explicit orders, had taken the offensive at the beginning of April. Catching the British off balance, he drove them out of Libya, invested Tobruk which he was unable to eliminate in his first assault, and finally came to a halt at Sollum on the far side of the Egyptian frontier. However brilliant tactically, Rommel's lightning advance was a strategic blunder. It failed to bring decisive victory, while adding another 700 miles to his already extended line of communications. As OKH had predicted, the resulting burden proved too much for his rear services to carry. In mid-May Rommel started complaining about his supplies for the first, but not the last, time.[25] His difficulties were not due, as has frequently been maintained, to the failure to put Malta out of action. Even in May, a peak month for the year so far, no more than nine per cent of the supplies embarked were lost *en route* to Africa.[26] From February to May, Rommel and his Italian allies received a total of 325,000 tons of supplies, or 45,000 more than current consumption.[27] Once he had started his offensive, however, the means at his disposal simply did not allow him to effectively bridge the enormous gap from Tripoli to the front. The result was that supplies piled up on the wharves while shortages arose in the front line. The quantities involved were often very small, e.g. a few tons of anti-tank ammunition that was urgently needed on 6 May,[28] but nevertheless significant. At the same time the Italians were experiencing even greater difficulties, because for 225,000 men they only had 7,000 trucks.[29]

June, therefore, was a month of crisis. Although a record quantity of supplies – 125,000 tons – were unloaded, the situation was 'in great danger every day' and Rommel was forced to live from hand to mouth.[30] On 4 April the Axis had re-occupied the port of Benghazi, only 300 miles from the Egyptian frontier, but this did not greatly improve matters, since only enough coastal shipping to carry 15,000 tons a month, instead of the projected 50,000, was available.[31] Though theoretically capable of processing 2,700 tons a day, Benghazi was well within reach of the RAF and suffered accordingly.[32] With an unloading capacity of 700–800 tons only exceptionally reached,[33] supplies continued to pile up at Tripoli while Rommel's situation became increasingly acute.

By his tempestuous advance the German commander had put himself in an impossible position. With Benghazi's capacity so limited, to stay where he was spelled certain disaster. To retreat was tantamount to admitting that OKH, where he was now known as 'that soldier gone mad',[34] had been right all along. The only way out of the predicament was to attack and capture the port of Tobruk. However, Rommel had to concede that his requirements for such an operation would be no less than four German armoured divisions – precisely the number originally envisaged by von Thoma.[35] This, however, was an impossible demand. Not only were Germany's forces now fully committed against Russia, but to grant Rommel's request meant that DAK would need another 20,000 tons a month, for which unloading facilities were not available.[36] The Italians were consulted and it was agreed that Rommel would have to make do with the forces at hand.[37] By way of consolation, the Axis units in Libya were, on 31 July, renamed *Panzerarmee Afrika* and Rommel, now elevated to full General, was put in command of both the German and Italian troops.

Whether, in view of what was to happen in 1942, even the capture of Tobruk would have helped Rommel very much is doubtful. The port was theoretically capable of unloading 1,500 tons a day, but in practice rarely exceeded 600. When consulted about its use, the German navy dismissed it as a disembarkation port for large ships, and bluntly told OKH that it would do well to rely exclusively on Tripoli and Benghazi to keep Rommel supplied.[38] Since in this period (July–August 1941) insufficient coastal shipping was available even to utilize Benghazi fully,[39] Rommel's scheme for solving his supply difficulties by capturing Tobruk seems highly impracticable.

Meanwhile, behind him, the situation in the Mediterranean was deteriorating. Early in June most of the German 10 Air

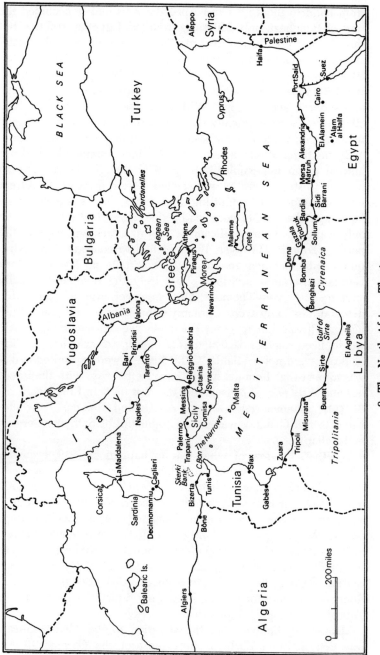

8. The North African Theatre

Corps, which had hitherto protected the African convoys from bases in Sicily, was transferred to Greece, so that the British aero-naval forces based on Malta and elsewhere gradually recovered much freedom of action. Losses at sea, which had hitherto been negligible, began to rise alarmingly. In July, nineteen per cent of all supplies (by weight) sent to Libya were sunk; in August nine per cent; in September twenty-five per cent; in October twenty-three per cent again.[40] In addition, Benghazi was heavily bombed in September, causing ships to be diverted to Tripoli and thereby extending the line of communications from 250 to about 1,000 miles. The various headquarters involved now began to blame each other. Rommel, always strongly anti-Italian, accused the Commando Supremo of inefficiency, demanding that the entire supply organization be taken over by the Wehrmacht.[41] The German Navy agreed, voicing the suspicion that the marked Italian preference for Tripoli might have something to do with their alleged desire to save their merchant fleet 'for the period after the war'.[42] OKH prepared a detailed study to show that the Luftwaffe had neglected the protection of the convoys in favour of attacks on targets in the eastern Mediterranean.[43] The possibility of sending supplies from Greece direct to Cyrenaica was studied, but this would have entailed dependence on a single-track railway from Belgrade to Nish, which was constantly being blown up.[44] On their side, the Italians argued that continued use of Tripoli was needed 'to split the enemy forces', claimed that they had no fuel-oil to enable their navy to deal with the Malta-based Force K, and demanded that the German air force tackle the job. However, when the head of OKW offered them German navy personnel to help operate the Libyan ports, his proposal was politely rejected.[45] Early in October, a half-hearted attempt was made to supply *Panzerarmee*'s most urgent needs by air, but this failed owing to a shortage of aircraft,[46] leading to more accusations and counter-accusations. At one point Rommel so far lost control of himself that he began to see imaginary British convoys passing through the Mediterranean, which earned him a sharp rebuke from the Armed Forces High Command.[47]

In all the confusion, one fact was entirely overlooked. Despite everything, the Italians succeeded in putting an average of 72,000 tons – or just above Rommel's current consumption – across the Mediterranean in each one of the four months from July to October.[48] Rommel's difficulties, therefore, stemmed less from a dearth of supplies from Europe than from the impossible length of his line of communications inside Africa. Thus, for example, the German commander discovered that he needed fully ten per cent of his precious fuel simply to transport the other ninety per cent.[49] If *Panzerarmee*'s fuel is put at about one third of its total requirements (excluding water and personnel), then it would be a reasonable guess that thirty to fifty per cent of all the fuel landed in North Africa was wasted between Tripoli and the front. Obliged to cover 1,000 miles of desert each way, thirty-five per cent of the vehicles were constantly out of repair. Under such conditions, any supply service was bound to break down.

November brought the inevitable crisis. During the night of the 9th an entire convoy of five ships carrying 20,000 tons was sunk off Cape Bon by British surface units, after which the Italians declared Tripoli to be 'practically blockaded'.[50] Supplies disembarked during the month dropped to a disastrous 30,000 tons,[51] while shipping losses rose to thirty per cent.[52] Since, however, Rommel's main fighting force consisted of his two German divisions, which together consumed about 20,000 tons a month and still had some supplies available, this was of less immediate significance than the fact that the British offensive which opened on 18 November made the routes inside Africa unsafe. British aircraft and armoured cars inflicted heavy losses on the lorry columns, simultaneously reducing their capacity by half, by restricting movement to night-time only.[53] For a few days after 22 November both divisions were actually cut off and only an occasional convoy got through.[54] Under these circumstances the front could not be held, and on 4 December Rommel ordered a general retreat. Curiously enough, the diary of the DAK quartermaster for the same day reads: 'Supply situation favourable from every point of view.'[55]

Initially, retreat made the situation more difficult still. Not only did the coastal road become clogged with westward-moving traffic, but the need to evacuate stores made the shortage of vehicles even more acute. Since there were no tanks to escort them, fifty per cent of *Panzerarmee*'s lorry columns were shot to pieces by British armoured cars.[56] But the retreat dramatically cut the distances to be covered and on 16 December, with Rommel near Benghazi and preparing to evacuate the town, we are told that 'Ib [quartermaster] officers [of DAK] have no worries; the divisions are fully supplied.'[57]

Panzerarmee as a whole, however, had far less cause for satisfaction. On 14 November, German pressure forced the Italians to resume convoys to Tripoli,[58] but initially this led only to further losses. Fuel in particular was so short that, by the middle of December, the Luftwaffe in Africa found itself limited to a single sortie per day.[59] This, of course, could not be allowed to continue. Overriding his admirals' objections, Hitler resolved to send German U-boats into the Mediterranean, and the imminent reinforcement of the German air force in the Mediterranean by units coming from Russia was announced on 5 December.[60] As an interim measure, they brought heavy pressure to bear on the French to sell them 3,600 tons of fuel.[61]

Meanwhile, the Italians made an all-out effort to save Rommel. Having materially assisted his retreat by using warships and submarines to bring fuel to Derna and Benghazi, they next made a supreme effort and sent four battleships, three light cruisers and twenty destroyers to escort a convoy to Libya on 16–17 December. Although the operation was successful – only the battleship *Littorio* was damaged – it dramatically highlighted another of the problems afflicting the Axis presence in North Africa. Owing to the extremely limited capacity of the ports, only four ships could actually be escorted, and even so one of them had to leave the convoy and head for Benghazi rather than Tripoli.[62] With 100,000 tons of warships being used to protect 20,000 tons of merchant shipping, the cost in fuel became prohibitive. Only once, at the beginning of January, could an operation of these proportions be repeated.

Mussolini was making use of the crisis in order to revive his cherished designs on Tunisia. On 2–3 December he suggested various measures (including the supply of German oil to the Italian Navy and the use 'on a grand scale' of German air transport) to alleviate the situation, but at the same time added that only Bizerta could definitely solve the problem.[63] Hitler agreed to the first two measures, but feared that excessive pressure might drive the French into the British camp, and therefore turned down the Duce's Tunisian plans, adding that the route from Bizerta to Libya was too long in any case.[64] On 8 December the chief of the Italian Armed Forces High Command accordingly considered the matter 'liquidated'.[65] Once Rommel had evacuated Benghazi on 24 December, however, Mussolini returned to the charge, demanded that Germany make 'concessions' to

France, and said he was ready to use force, even 'the entire Italian navy', if necessary.[66] A horrified Hitler allowed negotiations with Vichy to be reopened, but these led to nothing.[67]

Whether the use of Bizerta would have helped Rommel very much is doubtful. The problem of port-capacity in Africa would have been solved at a stroke, but another 500 miles would have been added to his overextended lines of communication. Of these, over 300 miles were served by two separate railways, but the 150-mile gap from Gabès to Zuara would have had to be bridged by motor vehicles, of which there was a shortage. Bizerta itself, as well as the entire route to Tripoli, was well within striking range of Malta, and the railways would have made excellent targets for the RAF. In view of these facts it might have been preferable to employ 'the entire Italian navy' to capture Malta, though even this would not have removed the two basic problems afflicting Rommel – the capacity of the ports and the distances to be covered inside Africa.

Meanwhile, as the DAK quartermaster's diary shows, the situation was improving. Since only 39,000 tons got across the Mediterranean, it is clear that the improvement had little to do with any increased safety of the sea-routes. Rather, it resulted from the unexpected discovery of 13,000 tons of Italian fuel reserves near Tripoli.[68] Even more important, Rommel's retreat to El Agheila had reduced his lines of communication to a more manageable length of 460 miles. The arrival on 6 January 1942 of the second 'battleship convoy' with six vessels carrying supplies eased the situation still further. Though supplies reaching Africa during the month can hardly have exceeded 50,000 tons,[69] the new year found *Panzerarmee* and especially DAK feeling very much better.[70]

1942: *Annus Mirabilis*

The beginning of 1941 saw Hitler at the zenith of his power, dominating the whole of Western Europe and preparing to extend his hold over the Balkan peninsula also. One year later, though the territory under German control had expanded enormously, the situation had changed completely and already the shadows were gathering around the Third Reich. On 10 December 1941, the United States formally joined in the struggle against

Hitler. At the same time, the Wehrmacht in Russia was being held, then thrown back for the first time since the beginning of the war. Whether it is true, as some authorities argue, that Germany had already forfeited any hope of emerging the victor may remain an open question. Certainly, the one chance which she still stood was to concentrate all available forces in a supreme effort to defeat the enemy in the East before those in the West could bring all their resources to bear, and therefore everything else should have been subordinated to this one aim.

Whether, under these circumstances, a fresh attack in North Africa was warranted is questionable. In January 1942, Rommel was still holding an area large enough to defend. The shorter line of communications had improved his supply situation, and the burden of maintaining a logistic organization over hundreds of miles of desert now rested on the enemy. Though the arrival of Kesselring's 2 Air Fleet had done much to ease the situation in the central Mediterranean, the supplies would only become really secure if a railway from Tripoli to the front were constructed.[71] Not even Rommel could overlook the fact that logistic difficulties were bound to reappear once the offensive was resumed. Accordingly, he demanded another 8,000 trucks for his supply columns. This was out of the question, since at the time all four German armoured groups operating in Russia could muster only 14,000 lorries between them.[72] Rejecting this request, OKH, Rintelen and even Mussolini warned Rommel that any fresh advance would again disrupt his supplies.[73] For the second time running, the German general disregarded the warning and on 29 January, after another lightning-stroke, his forces re-entered Benghazi where they were fortunate to recapture some of their own stocks. The rest of the story can be read in the diary of that paradigm of military efficiency, his own DAK. On 9 February, 100 per cent supplies could no longer be guaranteed to the troops, and by the next day tactical developments had overtaken logistical possibilities to such an extent that, because of the enormous distances and chronic lack of vehicles, no more ammunition was reaching the forward troops. On 12 February the DAK quartermaster angrily demanded an urgent interview with Rommel. On 13 February it was announced that the advance would stop at El Gazala, 900 miles from Tripoli.[74]

From mid-February to May, the voluminous correspondence between *Panzerarmee Afrika* and OKH is for the most part free from complaints about the logistic situation.[75] This is remarkable, for while reinforcements had raised Rommel's strength to ten divisions (three German, seven Italian) and, accordingly, increased his requirements to 100,000 tons,[76] he was in fact receiving an average of only 60,000 tons during each of these four months.[77] This was rather less than what a considerably smaller *Panzerarmee* had received during the difficult period of June–October 1941, and yet it enabled Rommel, first to take the offensive, and then to prepare another and even more spectacular one. These seemingly incomprehensible facts can be accounted for by the following considerations. First, the demand for 100,000 tons a month was much exaggerated; it corresponded to the needs of ten German divisions at full strength, whereas the forces actually under Rommel's command were much smaller.[78] Second, *Panzerarmee* was able to maintain itself at 900 miles from Tripoli because Benghazi, which had contributed little during the previous offensive, was now operating at full capacity.[79] As a result, the distance to be covered by approximately one third of his supplies was reduced to a still formidable, but ultimately manageable, 280 miles.

As there were no ports of any size east of Benghazi, however, a further advance was bound to end in a new crisis. Mussolini and his chief of the Armed Forces High Command, General Cavallero, realized this and demanded that a halt be called, but Rommel intended to wait only for reinforcements before resuming his drive to Tobruk, the Libyan–Egyptian frontier, and beyond. Asked by Commando Supremo how he intended to keep his army supplied, Rommel confessed that he did not know – the logistic services would somehow have to 'adapt themselves' to the tactical situation.[80] Having failed to dissuade Rommel, the Italians tried to prevent him from carrying out his plans by taking advantage of Hitler's wish to capture Malta, the focal point of the British position in the central Mediterranean, before mounting a fresh offensive in Africa.[81] By preparing a landing of such proportions that it could not possibly be carried out before the end of July, they hoped to force Rommel to postpone his offensive until the autumn.[82] The newly-appointed German CIC South, Field Marshal Kesselring, spotted this intention and tried to persuade them to take Malta by a *coup de main*.[83] The Italians, however, were obstinate, and the question was referred to the Axis leaders at their meeting of 29–30 April. Since Hitler supported the idea

of an early offensive – he considered Egypt 'ripe for revolution' –
a compromise solution was worked out. Late in May, Rommel
was to attack and capture Tobruk. He was not, however, to cross
the Egyptian frontier, and was ordered to terminate operations by
20 June, to enable the Luftwaffe to redeploy for the seizure of
Malta.[84]

During the last thirty years, the merits of this decision have
been endlessly debated. From a purely operational point of view,
it was probably correct. By denying the Cyrenaecan airfields to
the British, Malta was isolated.[85] The logistic problem, on the
other hand, was far more complex. Rommel's supplies, as we have
seen, depended less on Malta than on the capacity of the Libyan
ports and the distances to be covered inside Africa. While the
former problem could perhaps have been solved, the latter was
bound to remain because there were not enough vehicles available to bring DAK up to establishment.[86] Thus, an advance to
Sollum, while not producing any strategic victory, would merely
have added another 150 miles to his already overextended line of
communications, at a time when the Italians were using the
excuse of a shortage of fuel oil in order to divert a growing proportion of the German commander's oil back from Benghazi to
Tripoli.[87] With the capacity of Tobruk so limited, even the capture of Malta would not have solved Rommel's supply problem.
This had become clear in June of the previous year, when the
length of *Panzerarmee*'s communications had caused endless difficulties even though supplies were crossing the Mediterranean
practically unhindered.

The problem facing the Axis, then, was a double one – how to
guarantee the safety of the convoys from Italy, and how to secure
an adequate port at a reasonable distance behind the front. The
solution of both problems was essential for success; overcoming
either separately would not do. Given these facts, two courses
were open to the Axis. One was to adopt the Italian proposal –
Rommel should stay where he was and the Italians should
capture Malta in their own time. Assuming that oil for the Italian
navy could have been found, and given some extension of the port
of Benghazi, this would have enabled Rommel to hold out indefinitely and to prepare a large-scale attack on Egypt at some
later date. Alternatively, enough reinforcements – another two to
four German armoured divisions[88] – should have been brought
up, and sufficient stores accumulated, to enable Rommel to take

Alexandria in one swoop. This, as Halder had pointed out as early as 1940,[89] would have solved the problem of port-capacity once and for all, while at the same time turning Malta into a strategic backwater and making it possible to starve her out.[90]

With or without Malta in Axis hands, it is questionable whether an advance on Alexandria would have been practicable. Even if Hitler had the additional forces at his disposal, bringing them to Africa would have increased *Panzerarmee's* requirements to a point far beyond the combined capacity of Benghazi and Tripoli. This in turn would have made the accumulation of stores for an attack a hopeless task, while the number of vehicles required to transport stores inside Africa was far beyond the strictly limited resources of the Wehrmacht. Perhaps the only way to solve the problem would have been to rid *Panzerarmee* of its useless Italian ballast; this had been Thoma's demand in October 1940. Answering it, however, would have led to a different war. As it was, the capture of Malta might have enabled Rommel to hold out indefinitely at a reasonable distance from his bases. This, after all, had been the task originally assigned to him. Any attempt to do more, as Cavallero and Rintelen never tired of pointing out, was bound to be frustrated by the hard realities of the logistic situation.

Under these circumstances, the solution actually adopted proved to be the worst possible. On 26 May Rommel started his offensive. On 22 June he captured Tobruk and found the port intact.[91] The Axis, however, was in no condition to exploit this success. Though shipping losses in June had risen hardly at all as compared with May, lack of oil for the navy caused the tonnage plying the Africa route to fall by two thirds, while supplies disembarked dropped from 150,000 tons to a disastrous 32,000. The fuel shortage, moreover, forced the unloading of even this small amount not at Benghazi but at Tripoli.[92] This made Rommel's situation desperate. Unable to stay where he was, he had either to fall back or 'flee forward' in the hope of living off the enemy.[93] The protests of the Italian High Command, seconded by the German CIC South,[94] were of no avail. Proclaiming that the supplies captured at Tobruk would carry him to the Nile, Rommel was determined to go on and Hitler, who had never been very keen on the Malta operation, supported him.[95] The Axis hopes now hung on 2,000 vehicles, 5,000 tons of supplies and, above all, 1,400 tons of fuel captured at Tobruk,[96] but this was just not enough. After an advance of another 400 miles, the 'difficult supply situation', as well as exhaustion and stiffening resistance, brought *Panzerarmee* to a halt on 4 July.[97] As Rommel himself subsequently admitted, he was fortunate to be halted at this point. Had this not been the case, he might have arrived at Alexandria with two battalions and 30 tanks, and his line of communications longer still.[98]

Although brought to a stop at Alamein, Rommel had by no means given up. He still intended to resume the attack after a few days' recuperation.[99] However, the full impact of his long communications line now made itself felt. Of the 100,000 tons needed each month,[100] Tobruk – itself hundreds of miles behind the front – could handle barely 20,000. Lorries were in as short a supply as always, and attempts to use the British railway from Sollum resulted in only 300 tons per day being transported instead of 1,500 as planned.[101] What was worse, the port and the sea-routes leading to it were hopelessly exposed to the attacks of the Egypt-based RAF. Sending supply ships straight to Tobruk (or to the even smaller and more vulnerable ports of Bardia and Mersa Matruh) was difficult. On the other hand, unloading them at Benghazi or Tripoli, 800 and 1,300 miles behind the front respectively, involved impossible wastage and delay. Faced with this dilemma, Commando Supremo hesitated. In July, disregarding a storm of protest from *Panzerarmee*, the Italians opted to unload at Benghazi and Tripoli, with the result that although only 5 per cent of the shipping was lost and 91,000 tons put across, it took weeks for the supplies to reach the front.[102] Rommel himself saw the dilemma clearly enough, but, he insisted that the Italians send their ships directly into Tobruk,[103] with the result that in August losses rose fourfold and the quantity of supplies put across dropped to 51,000 tons.[104]

The lessons from these facts were clear enough. With or without Malta in Axis hands, the ships going to Tripoli and Benghazi usually got through, while those sent further east only served to turn Tobruk into the 'cemetery of the Italian navy'.[105] In mid-August, therefore, the Italians decided to disregard Rommel and to concentrate on Tripoli and Benghazi.[106] With this, *Panzer-armee's* position became desperate. To stay where it was meant suicide. Although he had only 8,000 out of 30,000 tons of fuel he claimed he needed for August, Rommel now decided to stake everything on a final attempt to break through to the Nile. He was supported by Kesselring, who promised more tankers for Tobruk. When these were sunk he said he would fly in 500 tons of fuel a day. Kesselring's planes failed to arrive and, having spent 10,000 tons of precious fuel, 'that lout of a Rommel' after four days of fierce fighting at 'Alam Halfa found himself back where he had started.[107]

Having tried and failed, Rommel now realized that the game was lost. He even began to consider withdrawal from Africa, but at this point Hitler intervened and forbade any retreat.[108] Disregarding *Panzerarmee's* protests, the Italians in September continued to concentrate on Tripoli and Benghazi[109] – with the result that supplies put across rose to 77,000 tons and dropped only slightly in October. The fact that they never again met Rommel's full requirements was not due to losses at sea. In September these dropped back to their level of July, and although they rose again in October[110] they still remained well below those of August.[111] Rommel's difficulties were due rather to the dramatic fall in the shipping tonnage plying the Africa route.[112] Whether this reflected a real shortage of ships, or was due to Italian reluctance to lose more of them, is difficult to say. According to the best available figures, out of 1,748,941 tons of shipping that Italy possessed in the Mediterranean in June 1940, 1,259,061 had been lost by the end of 1942. However, during the period in between, 582,302 tons were added in the form of German and German-captured ships, and another 300,000 or so from new construction, salvage, etc, so that the overall tonnage still available at the end of 1942 must have amounted to some 1,362,682 tons, or seventy-seven per cent of what it had been when Italy entered the war.[113] Moreover, it is significant that Cavallero could, as late as mid-October 1942, describe Italy's losses for the year as 'light'.[114] In any case, Rommel refused to believe that a shortage of shipping existed and accused the Italians of favouring their own troops.[115] As the situation grew worse, the bickering between the allies intensified, but to no avail. At the start of the battle of El Alamein, Rommel's troops were down to three basic loads of fuel – instead of the thirty or so which he claimed were needed in Africa – and eight to ten of ammunition. Since the railway from Tobruk was flooded, the transport situation was again said to be 'very difficult',[116] and in fact, 10,000 tons of supplies were still at Tobruk, from where it was impossible to bring them to the front.

Conclusion: supply and operations in Africa

After the war in North Africa was over, Rommel bitterly commented that, had he received but a fraction of the troops and supplies that Hitler poured into Tunisia in a hopeless attempt to hold it, he could have thrown the British out of Egypt many times over. This claim has since been echoed by many other writers. However, it ignores the fact that the Axis' presence in Africa had been put on an entirely different basis by Rommel's retreat and by the Allied landings in North West Africa. Having seized both Bizerta and Toulon, as well as the French merchant fleet, the Axis now possessed the means with which to send reinforcements to Africa at a rate *Panzerarmee* had never known. Even so, however, they did not succeed in maintaining them there for very long.

The lessons of the period of the Libyan campaigns proper seem clear. First, Rommel's supply difficulties were at all times due to the limited capacity of the North African ports, which not only determined the largest possible number of troops that could be maintained, but also restricted the size of convoys, making the business of escorting them impossibly expensive in terms of the fuel and shipping employed. Second, the importance usually attributed to the 'battle of the convoys' is grossly exaggerated. At no time, except perhaps November–December 1941, did the aero-naval struggle in the central Mediterranean play a decisive part in events in North Africa, and even then Rommel's difficulties were due as much to his impossibly long – and vulnerable – line of communications inside Africa as to losses at sea.[117] Third, the Axis decision of summer 1942 not to occupy Malta was of far less moment to the outcome of the struggle in North Africa than the fact that the port of Tobruk was so small and hopelessly exposed to the attacks of the RAF operating from Egypt.

More significant even than the above factors, however, were the distances that had to be overcome inside Africa. These were out of all proportion to those that the Wehrmacht had met in Europe, including Russia, and there was little motor transport available to bridge them. Coastal shipping was employed on some scale in 1942, it is true, but given the RAF's domination of the air its effect was limited because, the nearer to the front a port lay, the more exposed to attack from the air it became. Given these facts, Rintelen was right in pointing out that only a railway could solve the supply problem. This, after all, was part of the British solution. The Italians, however, never mobilized the resources for this purpose, nor did Rommel have the patience to wait for them.

That the reverses inflicted on Rommel during the summer and autumn of 1942 were due to the non-arrival of fuel from Italy, or to the fortuitous sinking of a disproportionately large number of vitally-important tankers, has frequently been maintained but is in fact without foundation. A detailed scrutiny of the list of ships sunk between 2 September and 23 October 1942 reveals that, out of a total of twenty-seven vessels, only two were tankers.[118] Also, the average quantity of fuel that Rommel received during the months July–October was actually slightly larger than that which he got during the halcyon days from February to June.[119] This suggests that his difficulties stemmed from the inability to transport the fuel inside Africa, rather than to any dearth of supplies from Europe. This impression is reinforced still further by the fact that, during the battle of El Alamein, no less than a third of *Panzerarmee*'s very limited stocks were still at Benghazi, many hundreds of miles behind the front.[120]

Finally, the often-heard claim that Hitler did not support Rommel sufficiently is not true. Rommel was given all the forces that could be supported in North Africa, and more, with the result that, as late as the end of August 1942, his intelligence officer estimated that *Panzerarmee* was actually superior to the British in the number of tanks and heavy artillery.[121] To support these forces he was given a complement of motor-trucks incomparably more generous than that of any other German formation of similar size and importance, and if the problem of securing *Panzerarmee*'s communications inside Africa was, as a result of the above-listed factors, never quite overcome, Rommel himself was largely to blame. Too late, he realized that:

> The first essential condition for an army to be able to stand the strain of battle is an adequate stock of weapons, petrol and ammunition. In fact, the battle is fought and decided by the quartermasters before the shooting begins. The bravest men can do nothing without guns, the guns nothing without plenty of ammunition; and neither guns nor ammunition are of much use in mobile warfare unless there are vehicles with sufficient petrol to haul them around. Maintenance must also approximate in quantity and quality to that available to the enemy.[122]

Given that the Wehrmacht was only partly motorized and un-supported by a really strong motor industry; that the political situation necessitated the carrying of much useless Italian ballast; that the capacity of the Libyan ports was so small, the distances to be mastered so vast; it seems clear that, for all Rommel's tactical brilliance, the problem of supplying an Axis force for an advance into the Middle East was insoluble. Under these circumstances, Hitler's original decision to send a force to defend a limited area in North Africa was correct. Rommel's repeated defiance of his orders and attempts to advance beyond a reasonable distance from his bases, however, was mistaken and should never have been tolerated.

War of the accountants

The pitfalls of planning

In the course of our inquiries so far, we have deliberately concentrated on a number of the most spectacular mobile campaigns of all time, and though some of these ended less happily than others, there does not appear to have existed any very clear connection between the amount of preparation involved in a campaign and its success or lack of it. For example, Marlborough's march to the Danube certainly entailed incomparably less administrative difficulties than did the least of Louvois' sieges – which was probably one reason why he undertook it in the first place. Two of Napoleon's most successful campaigns – those of 1805 and 1809 – were launched with almost no preparation at all, while although his war against Russia was prepared on a scale and with a thoroughness unequalled in the whole of previous history, this did not prevent it from becoming a monumental failure. In 1870, the Franco–Prussian War came as a surprise even though military preparations for it had been going on for years, with the result that there was little connection between the course of the campaign and Moltke's plans. Nothing daunted, a generation of staff officers before 1914 spent every minute – including even Christmas Day – planning the next war down to the last train axle, yet when war came it was totally unexpected, and the meticulously prepared plans led to nothing but failure. Twenty-five years later, the entire German invasion of Russia was nothing but a huge feat of improvisation with a preparation time of barely twelve months – not much in view of the magnitude of the problem. Rommel's African expedition was launched without so much as six-weeks' preparations and with no previous experience whatsoever, yet is commonly regarded as one of the most dazzling demonstrations of military skill ever.

Given the very short preparation times allowed to many of these campaigns, it is not surprising that they had to be conducted on a logistic shoestring. Napoleon in 1805 did not succeed in obtaining even one half of the wagons with which he intended to equip his army. In August 1914, the outbreak of war caught the Great General Staff in the middle of a major reorganization involving the complete overhaul of the gigantic railway deployment plan.[1] In 1940, the Wehrmacht's invincible *Panzer* divisions consisted, for the most part, of Mark I and II tanks – small machines that had never been intended for anything but training purposes. Though German equipment was not designed with desert warfare in mind, much of it proved more suitable to the North African theatre of operations than anything the British possessed.[2] Together with one modern authority, the men who prepared and conducted these campaigns might have said that 'the central problem of logistic planning' involves the differential in lead time.[3] While it might take years to design and perfect a new piece of equipment, and more years to mass-produce an existing one, operational – not to mention political – requirements can change in a matter of weeks or even days. In view of these facts, the greatest soldiers of all times knew there is a limit to the length to which planning can, or should, go, and those who did not recognize this were not normally the most successful.[4]

Shifting political constellations and changing operational situations have probably prevented the vast majority of past commanders from conducting their wars with anything resembling the number and type of resources they would ideally like to have. This has meant that commanders have needed certain personal qualities, such as adaptability, resourcefulness, ability to improvise, and – above all – determination. Without these qualities, even the most clear-sighted commander with the best analytical brain would be hardly superior to an adding machine. For him to exercise them, however, it is necessary to have a flexible staff and a system of command not made rigid by over-organization.

Exactly how large the 'brains' of an army should be in comparison with its body is not easy to say, but again it must be noted that there is no obvious correlation between this factor and success in the field. Thus, Napoleon's Imperial Headquarters expanded endlessly until it came to resemble a little army of 10,000 men. Much of this was useless ballast, however, for the

Emperor always did everything himself. The organization set up
by the elder Moltke did not number more than perhaps two
dozen officers, yet managed to win some of the most brilliant
victories in all history. Schlieffen's Great General Staff may have
been as perfect a machine as can ever be built out of fallible
human components, yet its strategy was doubtful and its logistics
probably even worse. In 1941, the Germans unhesitatingly put up
to 70 divisions under the command of a single Army Group with
no more than 800–900 working rooms at its disposal; which did not,
in spite of everything, prevent them from winning a series of stun-
ning triumphs. Rommel frequently managed with hardly any headquarters at all, either touring the battlefield in his 'Mamut' radio
truck or flying over it in his Fiesler Storch reconnaissance plane.

Seen against this background of ill-prepared, hurriedly
assembled forces, often only partly equipped with the wrong
hardware, and ordered to launch a campaign at a moment's
notice, the Allied Expeditionary Force invading France in June
1944 represented a triumph of foresight and organization. Here
was an army which was free to dictate the starting date of its own
battle whose every movement had been planned in detail for two
years on end, an army which, to an extent unprecedented before
or since, was able to select, design, develop, test and manufacture the hand-tailored equipment it claimed was needed for the
task at hand, and which was commanded by men who, perhaps
because they had behind them a number of large-scale seaborne
invasions in two world wars, insisted on making detailed provisions for the loading and unloading of the last jerrycan of fuel. In
short, an army which like no other in history relied on systematic
planning in order to prepare its operations and carry them out;[5]
and on whose structure of command a word ought therefore to be
said.

As constituted in September 1944, the Allied Expeditionary Force
in France was made up of 47 divisions, divided into three Army
Groups – two American, one British – with six Armies between
them. Overall command over these forces was in the hands of
Supreme Headquarters Allied Expeditionary Force (SHAEF), a
mixed Anglo–American organization which was itself divided
into an Advanced Section of vast proportions and a Rear Section
larger still. Coordination between these bodies, as well as
between them and the other armed services of both nations, was
achieved by means of an immense network of committees, boards

and liaison officers. To concentrate on the logistic services of the U.S. troops alone, there were in November 1943 no less than 562 officers and men engaged on planning. These were distributed between a rear and forward headquarters (known as COMZ and ADSEC respectively) 90 miles apart, so that it was necessary to set up 'a rapid courier service' to enable them to communicate.[6] The system of split headquarters had originally been created to solve the Communications Zone's problem of operating on both sides of the Channel, but it was retained long after everybody had moved to the Continent and led to friction between the two, particularly when ADSEC tried to break away from its parent organization. Both COMZ and ADSEC, moreover, were subordinated directly to SHAEF, with the result that General Bradley, the senior American ground commander in France, could ask for supplies to be divided between his Armies but was unable to order it. The 1944 army, to sum up, resembled a Brontosaurus, except that, in the case of the latter day monster, the brain was large out of all proportion with the body, instead of vice versa.

Apart from its structure, the thought-processes of 'the brain of an army' deserve comment. These can best be studied, not on the basis of some supposed 'national' characteristics but from the actual records produced by the instrument of command. Headed as it was by a man who thought the greatest military quality to be the ability 'to weigh calmly the factors involved in a problem and so reach a rock-like decision',[7] SHAEF in 1944 seems to have adopted Moltke's slogan *'Erst wegen, dann wagen'* (first consider, then risk) without, however, knowing how to strike the balance between its parts. As the records reveal, decisions – often on very minor issues – were made as follows. First, an order to work out some matter – e.g. whether a particular port should be made to handle rations or POL – was issued. Responsibility for this was then handed down the chain of command until, inevitably, it reached somebody who was no longer able to pass it on. A reconnaissance of the port in question would then be made, and a report written on every aspect of the question. Next, the memorandum would start its climb back through the hierarchy with every successive authority adding more arguments, reflecting a progressively wider point of view, until it finally reached the man in charge. On the basis of all the information thus accumulated, he would then announce his decision. The process appears systematic, businesslike, even magnificently rational. To find out just how effective it was is the task of the rest of this chapter.

Before embarking on this task, however, it remains to note that, from the historian's point of view at least, the exercise of combing through the SHAEF documents is a disappointing and, for this very reason, a singularly instructive one. The amount of information provided in the bulky files is huge, the number of graphs, tables and statistics far in excess of anything found in the papers of other modern armies (e.g. the Wehrmacht in 1941), not to mention previous ones. Yet when all these masses of material have been sifted and digested, in the final analysis little of real informative value can be found. To say just why this is so will only be possible after the completion of our study. Meanwhile, it is worth noting that of all the campaigns studied in this book the one waged by the Allies in 1944–5 alone gave rise to the question whether, in view of the balance of forces involved, it was a walk-over.[8] To answer this question is not our purpose, but its very existence is surely indicative.

Normandy to the Seine

'War', Napoleon is said to have mused on the eve of the battle of Borodino, 'is a barbaric business in which victory goes to the side that knows how to concentrate the largest number of troops at the decisive spot.' Depending on one's point of view, identifying this point may be a matter either of genius or of sheer good luck. Once it is identified, however, the feeding into it of men and material is a question of bases, lines of communication, transport and organization – in a word, of logistics.

The men who planned operation 'Overlord' were well aware that the success of an eventual Allied invasion of Europe would depend above all on their ability to feed-in troops and equipment at a higher rate than the enemy. While the problem itself was not in principle different from those facing commanders of all ages, the Allies' approach to it was unique. Starting approximately eighteen months before the invasion, a huge theoretical model consisting of thousands of components was gradually built up, the aim of the exercise being to achieve a comprehensive view of *all* the factors that would affect the rate of flow.[9] After some months had been spent on the construction of this model, the most important of these factors were identified as follows:

(*a*) The number of landing craft, coasters, troop transports, cargo ships, tankers and lighters likely to be available on D day. These determined the maximum amount of troops and equipment that could be landed within a given time. Moreover, since they all had to return to base (wherever that may have been) in order to reload, it was vital that those bases should be as near as possible in order to reduce turning time.

(*b*) The size and number of the beaches, their gradient (a vital consideration if ships of all types were to come as close ashore as possible, thereby dispensing with complicated transfer-arrangements which would inevitably have become bottlenecks) as well as the prevailing conditions of tides, winds and waves. Access from the beaches to the areas further inland was vital. Since it was thought that the beaches, however well selected and operated, would not be able to meet all requirements for long, it was decided to supplement them by constructing two artificial harbours and towing them across the Channel in sections, which again meant that certain geomorphological and meteorological conditions had to be met.

(*c*) The availability at a reasonable distance from the beaches of deep water ports of considerable capacity (which itself depends on a whole series of interrelated factors[10]). It was assumed that only such ports could secure the Allied foothold on the Continent in the long run, especially as the landing craft and artificial harbours would present only a temporary solution until the advent of winter conditions made their operation impossible.

(*d*) The feasibility of providing air support, on which depended the Allies' ability to hamper enemy operations and slow down his attempts to bring up reinforcements.

Having constructed this model, the planners under General Morgan went to their maps and began looking for places in Europe where these conditions could be met. In doing this, they soon discovered that an 'ideal' landing site did not exist, as the conditions set in the model often contradicted each other. Thus, beaches that appeared ideal from the point of view of their gradient (as in some places along the Pas de Calais coast) did not allow easy access to the country beyond, being bound by high and extensive sandy dunes. The Bay of Biscay area afforded a number of good ports but was beyond the range of fighter support. In their attempts to examine the 3,000 miles of German-dominated coastline and arrive at the best possible combination of all the relevant factors, Morgan's men might well have kept looking forever were it not for two basic considerations. First, it was imperative to launch the invasion somewhere not too far away from the Allies' main base in Britain so as to allow the landing craft and other shipping to turn quickly, a condition which in itself ruled out both the Mediterranean and Norway, leaving only the western coasts of France. Second, the landing sites had to be within reach of the RAF's Spitfire planes operating from Britain. Taken together, these two factors limited the choice by about ninety per cent, leaving only north western France, either the Pas de Calais or Normandy.

From a strategic point of view, the Pas de Calais presented the shortest and most direct route into Germany, as well as the opportunity of cutting off all German forces south of the Seine by means of a bold thrust eastward to Paris and beyond. On the other hand, it was also the most heavily defended area, and one which could be quickly reinforced by means of an excellent road and railroad network. In finally selecting Normandy, the Allies settled on an area which, owing to its geographical configuration as a peninsula jutting westward into the Atlantic, was relatively easy to isolate from the rest of France. However, this very factor was equally likely to operate against them in their attempt to break out after obtaining their initial foothold.

The landing sites having been selected, a comprehensive plan for logistic support during the first ninety days was worked out. The plans laid down exactly how many troops would be landed, where, when, and in what order; the procedures for clearing and operating the beaches, as well as the places where dumps were to be established; the point at which one system of unloading would be replaced by another (e.g. the shift from landing craft and DUKWs operating over the beaches to Liberty ships coming from Britain and later to other deep-draft ships sailing directly from the USA); and even the moment when one method of packing would be substituted by another (e.g. the replacement of jerrycans by bulk POL on D+15). To make sure that all these hundreds of thousands of items would be unloaded at the right time and place, a rigid order of priorities was worked out, laying down detailed procedures for the storing, requisitioning, packing, forwarding and distributing of literally every single nut and bolt. Since the Allies intended to capture a whole series of ports and use them for unloading troops and supplies, restoration plans for more than a dozen of them – including even those capable of handling no more than a few hundred tons a day – were worked out, an attempt being made to determine in advance just what resources would be needed for the purpose, where, and when.[11] This planning was certainly comprehensive, and it is not surprising that it took all of two years to complete.

In view of the unparalleled scope and thoroughness of the plans, it might be expected that the victory was due largely to their successful implementation. This, however, did not prove to be the case. Within hours of the first landings, all plans for orderly unloading were brought to nought by the heavy surf and, especially in the American sector, in face of fierce enemy resistance. A navigating error led to the landings being carried out in the wrong place and in the wrong order, with the result that engineer units preceded the assault parties to the beaches and, considerably reduced in men and material, had to work without the latter's protection. For several days after D day progress in clearing the beaches was slow and not enough exit roads could be opened, with the result that the whole area became hopelessly congested, a perfect target for the Luftwaffe, had it been able to intervene. The waterproofing of many vehicles turned out to be inadequate and numbers of them were lost. Required to make voyages of up to ten or twelve miles in a choppy sea, overloaded DUKWs ran out of fuel and sank. Since lorries were in short supply, DUKWs travelled further inland than would otherwise have been the case, thus taking longer to turn and performing poorly. Under these circumstances, supplies unloaded on the beaches during the first week amounted to only one half of the planners' forecast. Shortages, particularly in ammunition, quickly developed and made it necessary to resort to rationing.

While many of the above difficulties were due to the hazards of combat, some resulted from faulty planning which, in its desire to ensure that all the components of the logistic machine would mesh perfectly with each other, was too rigid and too detailed. Thus, only 100 tons of shipping per day – rather less than one per cent of the supplies to be unloaded even as early as D+12 – were set aside for emergency purposes. Provisions for flying in

6,000lb. per day of emergency stores did exist, but there was no way to effect a resupply within less than forty-eight hours of requisitioning. As might have been expected, the attempt to adhere to a rigid order of priorities in unloading supplies on the Continent led to great confusion and interminable delay, with men in small boats bobbing from one ship to another in order to ascertain the nature of their cargoes. The timetables governing the reloading of ships and landing craft upon their return to England were too tight and did not allow for delays, with the result that ports became congested and had to be cleared by emergency measures, in the course of which many a unit was torn apart or simply lost. Confusion in the ports was increased still further by the fact that no less than three organizations – MOVCO, TURCO, and EMBARCO – were responsible for embarkation. But the plans' greatest fault was that they did not allow sufficiently for the inevitable friction of war. Intended to eliminate waste, their tightness actually caused it because difficulties at the lower end of the pipeline would immediately reverberate all along its length.

Beginning on D+1, a start was made on the construction of the two artificial harbours. Highly complex and extremely costly to build and tow across the Channel, these installations failed to meet expectations. The British one in particular reached Normandy only after forty per cent of its component parts had been lost at sea, with the result that performance fell far behind expectations.[12] The American artificial harbour arrived more or less intact, but had hardly started operating when it was swamped by a gale and literally blown to pieces. In the event, it was the time-honoured device of blockships that proved most useful in providing protection for the landings. Much of the complicated and expensive equipment of the artificial harbours (some of which had been designed for entirely different operations and was apparently taken along simply because it was available) only served to clutter the landing area with wreckage and thus increased the danger to shipping. Here, as elsewhere, the men who planned operation 'Overlord' had clearly violated Hindenburg's maxim that, in war, only the simple succeeds.

While beaches and artificial harbours had to be relied upon to support the early stages of the operation, its ultimate success could only be assured if deep water ports were captured and restored to use. Since tactical developments on the Continent proceeded more slowly than had been anticipated, however, these hopes were at first disappointed. Thus, Cherbourg – by far the most important port in Normandy – began operating six weeks behind schedule and it was several more weeks before it reached its scheduled capacity of 6,000 tons per day. Saint-Lo, supposed to fall by D+9, did not in fact do so until D+48. Granville and Saint-Malo were expected to start operating around D+27 but remained in German hands until D+50. The proximity of the Normandy landing sites to the Brittany ports, Brest in particular, had played a major role in their selection for 'Overlord'. In the event, these ports were only taken months behind schedule and by then proved too far away from the front to be of any use. Other ports, such as Grandchamp and Isigny, were occupied more or less on time but were too small to make any significant contributions to Allied supply. As a result of all these factors, total American supplies disembarked during the month of June reached only 71 per cent of program, which led to a whole series of operations – 'Axehead', 'Lucky Strike', 'Beneficiary', 'Hands Up', 'Swordhilt' – being successively initiated, considered and rejected for purely logistic reasons.[13]

If the above facts did not lead to the collapse of the Allied operations in France, this was due to the fact that the beaches, even though deprived of the protection that was supposed to be afforded by the artificial harbours, proved capable of discharging supplies far in excess of what the plans had allowed for. This, however, was achieved only by disregarding all the plans. The first breakthrough came on D+2 when it was decided to ignore the predetermined order of priorities and unload everything regardless. Having for months resisted the beaching of its craft during low tide (a fact, incidentally, that had done much to determine the exact date and hour of the operation) the Navy suddenly discovered that such a procedure was feasible after all, thereby making it possible to discharge cargoes directly to the shore and dispense with many of the laboriously assembled boats, floating piers, pontoon causeways and the like. Thus, unloading proceeded not so much in accordance with the plans as without them, and in some cases against them, which was additional proof that the planning staffs had grossly overestimated the value of complex artificial arrangements and underestimated what determination, common sense and improvisation could achieve.

Another factor which prevented the shortcomings of the logistic apparatus from exercising too negative an influence was the planners' overestimate of consumption. This resulted partly from the fact that tactical progress was much slower than expected – by D+19 the Allied lodgement covered only about ten per cent of its anticipated area. With distances limited to no more than a few dozen miles in any direction, vehicles consumed only a fraction of the POL provided for them. How inaccurate the planners were is illustrated by the fact that, although unloading on the Continent continued to lag somewhat behind expectations throughout the months of June and July, it actually proved possible to land additional divisions *ahead* of schedule from D+24 onward.

If the slow unfolding of tactical operations reduced consumption of some supplies, it also caused problems in other fields. Thus, expenditure of ammunition – particularly small arms, hand grenades and mortar bombs – in the Normandy *bocage* was heavier than expected, as were losses in ordnance. Storage space was hard to come by so that camouflage and dispersion often had to be dispensed with. The volume of vehicular traffic in the shallow lodgement caused congestion and was soon to lead to the deterioration of the relatively few roads available. Though the railway network in Normandy was captured, against expectations, 'almost intact'[14] rail-traffic was uneconomical owing to the small distances involved. The fact that the front was only some twenty miles away from the beaches gave rise to friction between COMZ and the Armies, since the latter were naturally reluctant to relinquish control over their supplies in favour of the former. Friction also arose between COMZ and ADSEC, with the result that the former's transfer to the Continent was carried out ahead of schedule.

While it is therefore true that the slow unfolding of tactical operations in June and July 1944 caused some logistic problems, the net effect nevertheless was to create a lodgement area that was packed with troops and supplies. On the eve of operation 'Cobra', the breakout at Avranches, there were confined within the space of 1,570 square miles 19 American and 17 British divisions totalling a million and a half men, with supplies for the former alone averaging 22,000 tons a day. Though almost 90 per cent of this tonnage was still being discharged over the beaches, it was already clear that Cherbourg, once opened, would easily exceed the cautious estimates of its capacity made before D day.[15] Owing to the small distances, reserves of POL reached unexpectedly high levels while the state of ammunition supply improved dramatically during the lull in the fighting that preceded the breakout.[16] Yet amidst this cornucopia Allied supply officers continued to worry, making gloomy forecasts of shortages that were expected to materialize in the fall (when the weather would make the unloading of supplies over the beaches impracticable) and thinking of fresh variants of plans whose purpose it was to acquire the Brittany ports after all.

Uncomfortable about the state of supplies in Normandy, General Lee of COMZ tended to regard the continuation of operations in France with greater pessimism still. As originally conceived, the logistic plans for 'Overlord' had been based on the assumption that the Wehrmacht would fight a systematic defensive campaign, putting up resistance along one river line after the other. This in turn was expected to result in an Allied advance almost as slow and deliberate as that of 1918. While logistic support for this advance would have to be furnished mainly by motor transport (it was thought that up to seventy-five per cent of the railway network in France would be out of action due to either German demolitions or attacks by Allied aircraft) its rate would be sufficiently low to make it possible to leapfrog the railways from one river to the next. Supplies would thus proceed in an orderly manner from beach (or artificial harbour, or port) to railhead and from there to truckhead and Army dump. Operating in this way, the Allies hoped to reach the Seine by D+90 and the German border around D+360.

In the event, tactical developments assumed a very different form. By 25 July – D+49 – the Allies were still only holding the line they had hoped to reach on D+15. If the Seine was to be reached on schedule the distance between it and this line would therefore have to be covered in forty-one days instead of seventy-five as planned. A staff study, aimed at finding out whether such an acceleration was possible, was ordered by the SHAEF G–4 (supply officer), General Crawford, but concluded that it could not be done, for a deficit of no less than 127 quartermaster truck (GTR) companies was to be expected on D+90 and serious logistic difficulties encountered even as early as D+80. This was a very pessimistic view indeed of logistic capabilities.

Fortunately for the Allies, a decisive leader took charge of the situation. Unlike the majority of his fellow commanders, General Patton refused to be tied down by the logisticians' tables. Indeed, such was his indifference to them that, throughout the campaign of 1944–5, he only saw his G–4 twice – once before he assumed active command and again in the last week of the war.[17] Made operational on 1 August, Patton's 3. Army began its stormy career two days later by ignoring all plans and feeding no less than six divisions through the Avranches–Pontaubault bottleneck within seventy-two hours. At Pontaubault, they spread out like a fan and began advancing east and south-east, even though the single road serving as their artery of supply was still being counterattacked. By 6 August Patton's forces were threatening Laval and Le Mans. On 16 August they were in Orleans, and three days later the Seine was reached at Troyes. Moreover, Patton's advance out-flanked the positions from which the Germans had been trying to contain the Allied lodgement in Normandy and threatened to cut them off, with the result that, after the failure of their coun-terattack at Mortain, they rapidly fell back and thus enabled the remaining Allied forces, Hodges' 1. U.S. Army and Montgomery's 21. Army Group, to effect their own breakout and reach the Seine.[18] This they did, and the river's western banks were finally cleared on 24 August – fully eleven days ahead of the schedule which, according to the logisticians, could not be met. Mean-while, in a manner reminiscent of the proverbial kettle rattling behind the dog's tail, those same logisticians continued to insist that what Patton and Hodges were doing was impossible.

While it may be true that erring on the safe side is normally preferable to its opposite, the discrepancy between expectation and actual performance in this case is so great as to call for an explanation. Such an explanation is readily found when one realizes how modest were the demands that the planners made on the means at their disposal. Thus, for example, it was assumed that a truck company could provide no more than fifty miles forward lift a day (that is, overall distance covered in 24 hours was not supposed to exceed 100 miles[19]) whereas actual perfor-mance turned out to be at least thirty per cent higher.[20] Almost unbelievably, the plans had been made on the assumption that the state of the French road network would not allow supply by motor transport at more than 75 miles from the railheads[21] – a gross underestimate that was exceeded at least three or four times over. Finally, consumption by a single Allied division was put at 650 tons a day, whereas a force engaged in pursuit actually required only a fraction of this amount, possibly not more than 300 or 350.[22]

Even when these facts are taken into account, however, it seems difficult to reconcile the pessimism of the Allied planners with the superabundance of the means at their disposal. In late August 1944 there were twenty-two American divisions in France. Of these, sixteen were operating on or near the Seine at a distance of some 250 miles from Cherbourg whereas the remaining ones were still disembarking in Normandy or facing east towards Brittany. Putting their average distance from the ports at 200 miles and assuming that consumption did in fact amount to the regulation 650 tons, $22 \times 200 \times 650 = 2,860,000$ ton/miles forward lift per day were required. There is no evidence of just how much transport was available at this time, but even as early as 25 July it had totalled 227 GTR companies plus the equivalent of 108 more in rail transportation facilities.[23] Overall forward lift therefore amounted, even if only according to the regulations, to 3,350,000 ton/miles a day – and this does not take account of the fact that additional GTR companies were brought in during August, nor that the divisions were furnished with truck components so lavish that they were capable, as subsequent events were to demonstrate, of hauling a substantial fraction of their own supplies over hundreds of miles behind the Armies' rear boundaries.[24]

While the evidence is far from complete, the above facts suffice to show that the Allied advance from Normandy to the Seine, however successful and even spectacular strategically, was an exercise in logistic pusillanimity unparalleled in modern military history. This is even more astonishing when one compares the conditions prevailing during this campaign with those governing some of the others that we have studied. Not only were the Allies possessed of more motor transport than any other army had ever dreamt of, but they were operating in favourable summer weather and over a road network that was among the best and most dense in the world. There was little enemy air activity,[25] and a friendly population furnished considerable assistance rather than engaging in sabotage. Yet for all these advantages, the Allied offensive had to be carried out against the logisticians' advice. To paraphrase a famous quotation, in the kingdom of supply something was rotten.

'Broad front' or 'knifelike thrust'?

Of all the problems raised by the Allied campaign in north-western Europe in 1944–5, the question whether it would have been possible to put an early end to the war by means of a quick thrust from Belgium to the Ruhr is perhaps the most important. The literature on this subject is vast and still growing. Here we shall only attempt a brief summary of the main views. These appear to be as follows:

> (*a*) The one put forward by Chester Wilmot and, above all, Field Marshal Montgomery himself. These two have argued that the strategic opportunity presented itself in September 1944. Given a clear-cut decision and a willingness to stake out a meaningful order of priorities on the part of the supreme commander, 2. British and 1. American Armies could have captured the Ruhr – and perhaps Berlin as well. Eisenhower, however, refused to concentrate his logistic resources behind Dempsey and Hodges. In particular, he did not want to halt Patton's 3. Army. Consequently, the opportunity to end the war in 1944 was lost.[26]
>
> (*b*) To counter Montgomery's accusations, Eisenhower has defended his decision on strategic grounds, claiming that a thrust by part of his forces into 'the heart of Germany' was too risky and would have led to 'nothing but certain destruction'.[27] Subsequent writers have produced more arguments to justify his decision, including the need to avoid offending Allied (that is, American) public opinion by halting Patton, differences over the structure of the chain of command in France, and – last though not least – logistics.[28]
>
> (*c*) Finally, Basil Liddell Hart has suggested that, while the opportunity to go for the Ruhr did exist, failure to utilize it lay not so much with Eisenhower as with Montgomery himself. In particular, the discovery at the critical moment that 1,400 British-built lorries had defective engines was decisive. Given this, there was little that Eisenhower could do to help, for Patton's 3. Army was receiving so few supplies that not even by stopping it could enough transport have been made available to enable Montgomery to occupy the Ruhr.[29]

Although the accounts of the situation that faced the Allies early in September 1944 differ in conclusion and even in detail, there is little doubt that operations in the period just preceding had been among the most spectacular in history. Having sent its advance formations across the Seine on 20 August, Patton's 3. Army covered almost 200 miles in twelve days until it stopped in front of Metz. Hodges' 1. Army on his left advanced even further, reaching the Albert Canal in eastern Belgium on 6 September. Montgomery's 21. Army Group, whose rate of progress had hitherto been markedly slower than that of the American forces, now surpassed itself by surging forward across northern France and Belgium, capturing Antwerp – its port virtually intact – on 5 September and only coming to a halt on the Meuse–Escaut Canal four days later. It was a performance of which the originators of the *Blitzkrieg* would have been proud, and one which the Allies themselves had not foreseen.

As might have been expected in the light of previous events, all these operations were carried out against the advice of the SHAEF logisticians, who declared them to be utterly impossible. Compelled to revise their cautious estimates of July, they completed a new feasibility-study on 11 August. This showed that *if* a whole series of conditions was met it *might* be possible to support a tentative offensive by four U.S. divisions across the Seine on 7 September. Even this conclusion, however, was qualified by the suggestion that operations south of the Seine should be halted in favour of an attack on the Channel ports, and that the liberation of Paris should be postponed until late October when railways from the Normandy area would hopefully be available to carry relief supplies. As it was, Paris was liberated on 25 August. By the target date of 7 September both Patton and Hodges had already gone 200 miles behind the Seine. A week later 16 U.S. divisions were being supported, albeit inadequately, on or near the German frontier on both sides of the Ardennes while several more were engaged on active combat operations in Brittany. All this was achieved in spite of the fact that the conditions laid down in the paper of 11 August had been only partly met. Seldom can calculations by staff officers have proved so utterly wrong.

Supply over the rapidly expanding lines of communication – those in the American sector grew from 200 or 250 miles to over 400, and those in the British one from 80 to nearly 300 – could

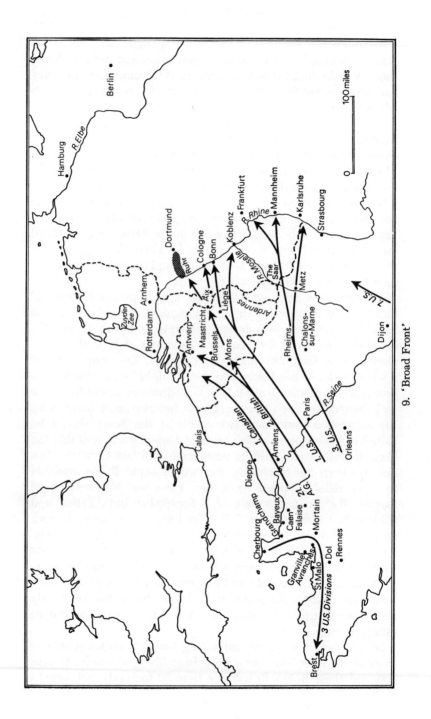

9. 'Broad Front'

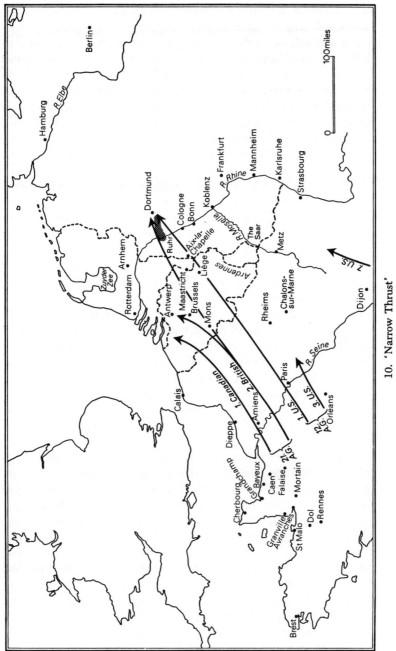

10. 'Narrow Thrust'

not, of course, be effected without abandoning all orderly procedures and employing emergency measures. With petrol in their
tanks and rations in their pockets, combat units, faced with very
little organized opposition, could accelerate their progress almost
at will. Not so COMZ, however, which found it impossible to
make depôts keep pace with the rapidly advancing front. No
sooner had a site been selected than it was left behind, and after
several attempts (each costly in terms of transport) COMZ gave
up in despair and concentrated on bringing up the most essential
items from bases that were sometimes as much as 300 miles in the
rear. The transport to carry the supplies across these distances
came from hundreds of units considered less essential – heavy
and anti-aircraft artillery, engineers, chemical warfare, and the
like – which were stripped of their vehicles and left immobilized,
sometimes on reduced rations, as were three divisions newly
arriving in France, whose organic trucks were taken away and
formed into improvised GTR companies. While rations, POL and
ammunition were being rushed forward, the supply of everything
else, including, in particular, clothing and engineering stores, had
to be deferred. Air transport was used on a grand scale but failed
to deliver more than an average of 1,000 tons a day because of
the lack of airfields near the front line and, even worse, because
aircraft were withdrawn to take part in a series of would-be
airborne operations that never materialized. The most famous of
all these expedients to supply the front was the Red Ball Express,
a loop system of one-way highways reserved exclusively for the
supply service over which thousands of trucks rumbled night and
day.

These emergency measures notwithstanding, the flow of supplies reaching the front gradually diminished until, on 2 September, 3. Army's advance was brought to a halt. 1. Army struggled
on for a few more days before it too came to a standstill. Compared with deliveries to both Armies of over 19,000 tons a day
during the second week of August, COMZ could only promise
7,000 at the end of the month, and it is doubtful whether even
this figure was in fact met.[30] Stocks in the hands of the front-line
units dwindled at an alarming rate, e.g. from 10.5 days of POL
supply at 1. Army on 5 August to 0.3 days on 2 September and
zero one week later.[31] Meanwhile, reserves available in Normandy
actually rose as they could not be transported forward. Though
consumption of ammunition had fallen off by thirty to ninety per

cent since July, deliveries fell so short of demands that a single corps (the XX) could request supplies in excess of all the tonnage allocated to 3. Army, to which it belonged, put together. Since whatever ammunition dumps that could be established were quickly left behind, both 1. and 3. Armies resorted to the creation of rolling magazines, a procedure which, though effective in ensuring that at least some supplies would always be available, wasted transport.

As always happens when supplies are short, friction and even tension arose between the front-line troops and the supporting services in their rear. 3. Army in particular was notorious for the unorthodox means it employed in order to obtain what it needed. Roving foraging parties impersonated members of other units, trains and convoys were diverted or hijacked, transportation companies were robbed of the fuel they needed for the return journey, and spotter planes were sent hundreds of miles to the rear in order to discover fuel-shipments. Inside the zone of communications itself, the strain on men and material working around the clock led to fatigue, accidents, malingering and occasional sabotage. Vehicles went without maintenance until they broke down and the number of major repairs needed rose rapidly. Record-keeping of the movement of supplies, which had always been a weak point in the Allies' organization, became even worse during the rapid advance of August, with the result that some of the limited transport available was wasted on items which the Armies had not requisitioned and did not need. Occasionally, instances of a worse kind of waste occurred. Thus, out of 22,000,000 jerrycans more than half had been lost by the end of August, with the result that this humble item limited the entire POL supply system. Supply discipline, especially at 3. Army, was poor and led to enormous quantities of equipment, particularly clothing, being left behind, so that the salvage companies were swamped with work.[32] Instead of trying to capture French locomotives and rolling-stock, Patton's men deliberately shot them up – finding them, as one author puts it, 'a rewarding target in more senses than one'.[33]

Though the logistic situation of the British forces operating in Belgium was also strained, it was not as difficult as that of the two American Armies. Distances here were smaller – 21. Army Group was operating on the 'inside track' along the coast – and roads were more numerous than in Lorraine. On 30 August, when

Hodges and Patton had run out of fuel, Dempsey informed
SHAEF that his supply situation was 'very favourable'.[34] Though
the length of their communications increased to almost 300 miles
during the subsequent week, the British coped by reducing unloading at the ports from 17,000 to 6,000 tons per day – thus
freeing truck companies – and immobilizing many units, including one complete corps (the VIII).[35] So rapid was the advance
that the Belgian railroad network, the world's best, was captured
intact, though its utilization was delayed because the Americans
did not hand over rolling-stock as scheduled.[36] Above all,
Montgomery's advance did not take him away from the ports, as
did that of Bradley's forces. On the contrary, it put him in a position to clear up a whole series of ports along the Channel – including Le Havre, Dieppe, Boulogne, Calais and Dunkirk – so that
their situation regarding unloading capacity promised to become
much better than had been expected even a month earlier.

Meanwhile, what of the German enemy? Since D day, the
Wehrmacht's casualties in the west had been staggering; 400,000
men, 1,800 tanks and assault guns, 1,500 other guns and 20,000
vehicles of all kinds had been lost.[37] The units that still remained
had been battered beyond recognition by Allied air and land
forces, had been compelled to withraw from a series of pockets
such as those at Falaise and Mons and had lost all resemblance
to organized fighting formations. Thus, for example, 1. SS division
escaped over the Seine with only 40 tanks and 1,000 fighting men
left. 84. infantry division had 3,000 men whereas 2. armoured division had only 2,000 men and 5 tanks; 2. parachute corps, despite
its name, numbered only some 4,000 men – hardly more than a
strong brigade – with all but light weapons lost.[38] According to
its own estimates, the Wehrmacht was outnumbered along the
entire front by ten to one in tanks, three to one in artillery (this
particular imbalance was regarded as 'not too serious' in itself,
but the Germans suffered from a shortage of ammunition) and
to an 'almost unlimited' extent in the air.[39] Hitler, moreover,
concentrated most of his forces on the Moselle in order to stop
Patton, with the result that, according to Eisenhower, he only
had the equivalent of two 'weak' armoured and nine infantry
divisions available north of the Ardennes, all of which were said
to be 'disorganized, in full retreat and unlikely to offer any appre-
ciable resistance'. Indeed, such was their state that a member of
Eisenhower's staff (probably COS Bedell Smith himself) con-

sidered it 'possible to reduce the strength of the Allies' forces in this area by 3 divisions and, far from prejudice the advance, aid it'.[40] In short, the route to the Ruhr – Germany's industrial heart, where more than half of her coal and steel were produced – lay wide open. As Liddell Hart has written, rarely in any war had there been such an opportunity.

In order to understand why this opportunity was missed, it is necessary to return to the spring of 1944 when the Allies' basic strategy for the coming Continental campaign was laid down. In a memorandum of 3 May – signed, incidentally, by British officers – it was laid down that Berlin, while forming the Allies' ultimate objective, was too far away and that their first goal in Germany should therefore be the Ruhr. While it was recognized that the easiest and most direct route to this area led north of the Ardennes through Liège and Aix-la-Chapelle, it was decided that there should be a second thrust through the Saar in order to compel the Wehrmacht to stretch its resources and keep it guessing as to the Allies' intentions. On 27 May, the principles formulated in this memorandum were embodied in a directive issued over Eisenhower's signature.[41]

Whether the best way to utilize an overwhelming numerical superiority such as the Allies possessed is to split it up so as to enable the enemy to operate on internal lines will not be discussed here. Suffice it to say that the battle for Normandy had, as we saw, developed in a manner very different from what had been expected. While progress had initially been slow, the breakthrough when it came proved much more decisive than anyone had hoped. In view of this, Field Marshal Montgomery began to entertain second thoughts as to the wisdom of the Allies' strategy and – so it seems – first voiced them to his American colleagues on 14 August.[42] Three days later, his ideas had crystallized sufficiently for him to tell Bradley that 'after crossing the Seine, 12. and 21. Army Groups should keep together as a solid mass of forty divisions which...should advance northwards...with their right flank on the Ardennes'.[43] Montgomery, however, did not see Eisenhower until 23 August, and by then had modified his argument. Instead of demanding that the whole of 12. American Army Group be deployed to protect his right flank in Belgium, he now considered that this task would have to be carried out by the nine divisions of Hodges' 1. Army alone. The rest of 12. Army Group – that is, Patton's 3. Army – was to be kept in its place, for the Allies did not possess enough logistic resources to support a simultaneous advance by all their forces and, in trying to be strong everywhere, would end up by being too weak to gain a decisive victory at any one point.[44] While Eisenhower refused to halt Patton – instead, he authorized him to continue eastward at least as far as Rheims and Châlons-sur-Marne – Montgomery gained his point in that Hodges' entire Army was ordered to advance north of the Ardennes in conjunction with 21. Army Group.[45] However, Eisenhower intended that the main objective of the northward thrust would not be the Liège–Aix-la-Chapelle gap, but the port of Antwerp, without which he believed no further advance into Germany could be sustained.[46]

With that, the first round of the great argument came to an end. It took place at a time when Montgomery was only just completing his eighty-mile approach to the Seine, whereas Patton, who had far longer distances to go, was already across the river and advancing fast. Apart from all considerations of national prestige and public opinion, it is not surprising that Eisenhower did not want to stop his most dashing commander in favour of one who, up to that point, had scarcely proved himself a master of pursuit operations. Ten days later, however, 21. Army Group had pushed forward 200 miles and was closing on the German border. With his position thus immensely strengthened, Montgomery wrote Eisenhower that:[47]

1. I consider that we have now reached a stage where one really powerful and full-blooded thrust towards Berlin is likely to get there and thus end the German war.
2. We have not enough maintenance resources for two full-blooded thrusts.
3. The selected thrust must have all the maintenance resources it needs without any qualification and any other operation must do the best it can with what is left over.
4. There are only two possible thrusts: one via the Ruhr and the other via the Saar.
5. In my opinion the thrust likely to give the best and quickest results is the northern one via the Ruhr.

The wording of this message was extremely unfortunate, for the objective mentioned – Berlin – was still over 400 miles away and, as we shall see, definitely beyond the reach of the Allied forces at this time. Nor did Montgomery aid his cause by sending, a few days later, an 'amplification' in which he spoke of a 'knifelike thrust...into the centre of Germany'.[48] These exaggerations subsequently enabled Eisenhower to claim, with some justification, that the British Field Marshal's proposals were reckless and ill-considered.

The confusion as to the nature and direction of the proposed offensive in the north increased still further when, on 8 September, the first V 2 landed on London and caused the British government to demand that 21. Army Group overrun the launching sites in Holland. This led to Montgomery modifying his ideas once more, and instead of striking east at the Ruhr he now decided to move north to the Zuider Zee. In so far as he authorized the Arnhem operation, Eisenhower supported him. However, he remained adamant throughout that no advance into Germany was possible unless Antwerp was first opened up.[49]

Thus, Montgomery's own ideas as to just what he wanted to do were much less clear than he would have us believe. His original 'solid mass' had been whittled down to a 'knifelike thrust', and his mention of Berlin as the objective continued to bedevil all discussion.[50] The confusion surrounding the issue, however, should not be allowed to obscure the main question, namely whether it was logistically possible, in September 1944, to capture the Ruhr *without first opening Antwerp*. The argument as to whether or not Germany was at that time on the point of collapse[51] is not directly relevant to this question, for there is no doubt that, whatever forces Hitler was or was not able to mobilize 'in the centre of Germany',[52] the way to the Ruhr lay wide open in the early days of September. The only question, therefore, is whether the logistic means at the Allies' disposal were sufficient to support a drive for the Ruhr.

First, a few definitions must be made. For the purpose of our inquiries, a 'drive to the Ruhr' is taken to mean an advance by 2. British and 1. American Armies, with a combined total of 18 divisions, to Dortmund, a move which would have left the Ruhr encircled. The starting point of the British forces would have been the Meuse–Escaut Canal (that is, if the Arnhem operation had not been launched), whereas the Americans would have set out from the Maastricht–Liège area. The distance from these starting points to the objective of Dortmund is almost equal and amounts to some 130 miles. It is assumed that all railway transportation facilities in Germany would have been put out of action and that no supply by air to the forward troops would have been possible owing to the non-availability of airfields. All calculations are made in terms of 200-ton American GTR companies, which was the practice of the SHAEF logisticians themselves. The target date for the start of the attack is taken as 15 September.

If a consumption of 650 tons per division per day is assumed (as an average for both American and British units, this may be somewhat too high) overall supplies required for the operation would have amounted to $18 \times 650 = 11,700$ tons a day. Of this, the British with their 9 divisions would have accounted for 5,850 tons. This quantity would have been brought forward to the starting bases in Belgium by rail (2,800 tons); by air, in planes diverted from the Arnhem operations, to Brussels (1,000 tons), and by road, in American trucks taken from three divisions immobilized in Normandy (500 tons), making a total of 4,300 tons.[53] The remaining 1,550 tons Montgomery would have had to bring up from the Caen–Bayeux area by means of his own motor transport. Assuming that each truck company was capable of no more than the regulation 100 miles per day, forty-six such companies would have been required for the purpose. The operation itself, assuming a turning time of two days for the distance of 260 miles[54] to Dortmund and back, would have required $(5,850 \times 2):200 = 58$ GTR companies in support. Since the British at that time possessed a grand total of 140 such companies,[55] thirty-six would have been left to supply the six divisions of 1. Canadian Army and for port clearance. It could be done – though only just.

By contrast, the situation of Hodges' 1. Army was considerably more difficult. His nine divisions consumed 5,850 tons per day, but his allocation at this time amounted to 3,500 only – and COM Z found it difficult to deliver even this. To bring up the difference from the railways south of the Seine, 200 miles away, approximately thirty-five additional GTR companies would have been required. The advance on Dortmund itself would have necessitated fifty-eight such companies, so that the total deficit amounted to 93.

The question, then, is how many lorries could have been freed if 3. Army, instead of being allowed to continue to the Moselle, had been halted on the general line Paris–Orleans. No detailed figures as to the amount of transport supporting Patton are available, but a rough estimate can be made. On 15 September, 3. Army was receiving at least 3,500 tons per day – possibly somewhat more, for Patton used his own trucks for hauling the line of communication, instead of confining them to the zone of operations. To move this quantity over the 180 miles between the above-mentioned line and his front, the equivalent – in one form or another – of fifty-two GTR companies must have been employed.[56]

In the middle of September, a U.S. corps – the VIII – was operating against Brest. Originally meant to secure adequate port capacity, this operation had by now become out of date. If it was nevertheless carried out, the reason was simply prestige.[57] Though detailed figures of consumption and transport for this operation are, as usual, not available, a rough guess can be made. At that time, the Allied-operated railway in Normandy reached from Cherbourg to the Dol–Rennes area. To bring up supplies for three divisions over the 80 miles from there to Brest, some fifteen GTR companies must have been kept busy.

In addition, there were the organic trucks of all these units. Though no detailed figures are available, there is reason to think that, if formed into provisional transportation companies, the vehicles of the nine divisions involved could have increased the quantity of supplies reaching Hodges in Belgium by another 1,500 tons.[58] His overall requirements would thereby have been cut by twenty-two GTR companies, and of the seventy-one such companies needed to carry out the drive on Dortmund $52+15=65$ would have been available. Our conclusion, though necessarily tentative,[59] must therefore be that it could have been done, though only just.

Conclusions

On the basis of the best information available, we have calculated that, in September 1944, enough transport could have been found to carry Dempsey and Hodges to the Ruhr. The number of truck companies at hand, however, is only one factor among the many involved. To complete our investigation, these must now be briefly considered.

Had Eisenhower accepted Montgomery's proposals and concentrated all his logistic resources behind the thrust by 2. British and 1. American Armies, twelve out of the forty-three Allied divisions in France would have been completely immobilized and their truck complements taken away. In addition, seven divisions coming up from the Mediterranean in operation 'Dragoon' were irrelevant to the main effort, so that only twenty-four would have been left over. Of these, six belonging to 1. Canadian Army would have been operating against the Channel ports. For the advance on the Ruhr proper, only eighteen divisions would have been available – a fairly small number, admittedly, but one which would in all probability have sufficed to break through the weak German opposition at this time.

Since distances to be covered inside Germany were relatively small, providing air support for the exposed flanks should not have been unduly difficult – as it would no doubt have been in an advance on Berlin. Instead of having to cross four rivers, which was what the Allies tried to do during the Arnhem operation, they would have found their way barred by two only. The road network leading eastward to the Ruhr was excellent and superior to that leading north into Holland.

Since transport to set up substantial forward dumps in Belgium, in addition to daily consumption, was not available, the drive on the Ruhr would have to be launched with the Allies' main supply base still hundreds of miles in the rear. This would no doubt have entailed long delays between the requisitioning of supplies and their arrival at the front, a fact that could have serious consequences in case of a sudden emergency. However, an allocation of 650 tons per day was generous almost to a fault, and Allied communications in France and Belgium – especially rail transportation facilities[60] – were being rapidly and daily improved. In view of this, the risk of operating at such distances from base could have been accepted, especially as the Luftwaffe was powerless to interfere with the line of communications.

Against this, it has been argued that the Allies, having already twice preferred strategic to logistic considerations during their advance across France,[61] could not have done so a third time without stretching their communications to the breaking point. This view, however, ignores the fact that the distance from Bayeux to Dortmund is no longer than that which Patton covered from Cherbourg to Metz. In other words, the Allies *did* succeed in supporting their forces over 450 miles – but did so in the wrong direction.

An interesting aspect of the problem is the fact that, contrary to what is generally believed, acceptance of the plan for a drive to the Ruhr would *not* have entailed the transfer of transport from American to British units, with all its consequent complications.[62] Rather, it was a question of switching vehicles from one American Army to another. Of the three Allied commanders in Belgium – Crerar, Dempsey, and Hodges – it was Hodges whose supply difficulties did most to prejudice Montgomery's plans.[63] Had Patton been stopped, it was to his 1. Army that the transport thereby freed would have gone. By refusing to halt 3. Army Eisenhower made his prophecy come true that no advance into Germany would be possible without Antwerp, for it was only when that port was opened in late November that distances behind Hodges were cut from over 400 miles to seventy. The British did not need Antwerp, as they were by that time living comfortably from the Seine and Channel ports.[64]

If Eisenhower must therefore be held responsible for failing to realize where the centre of gravity lay and adjusting his priorities accordingly, it cannot be denied that to have done so would have required almost superhuman foresight. At the time when the idea of a change in the Allies' fundamental strategy was first voiced, Patton was marching at full steam whereas Montgomery – even though the distances he had to cover were much smaller – was merely inching his way forward towards the Seine. By no stretch of the imagination was it possible to foresee that, during the next two weeks, this normally so cautious commander would suddenly surpass himself and advance 200 miles to the German border. Nor, it will be remembered, did Montgomery at first even ask that Patton be halted. On the contrary, he wanted him to protect the flank of his own offensive, and for that purpose demanded that the direction of Patton's advance be altered from east into Lorraine to northeast into Belgium. By the time he realized that forty divisions carrying out 'the Schlieffen Plan in reverse' could not be supported logistically it was already too late, as he himself admitted.[65]

Montgomery did even more to prejudice acceptance of his plan by engaging in loose talk, first of an advance on Berlin and then of a 'knifelike thrust' into the heart of Germany. While there is good reason to believe that a drive by eighteen divisions against the Ruhr would probably have succeeded, there could be no question of occupying the greater part of Germany with such a small force. As for knifelike thrusts, they may be successful if conducted over a distance of a mere 130 miles but become distinctly risky and even reckless when launched in search of objectives 400 miles away – and with the nearest base another 300 miles to the rear. Nor could such a thrust by even eighteen divisions have been supported logistically. Rather, Montgomery's staff were thinking in terms of twelve only. The latter's maintenance would have to be reduced to 400 (in some cases, 300) tons per day and even this was more than 21. Army Group could manage with its own resources.[66] As Eisenhower has rightly said, nothing but certain destruction could have resulted from such a move.

In the final account, the question as to whether Montgomery's plan presented a real alternative to Eisenhower's strategy must be answered in the negative. Though our computations seem to show that the means required for a drive to the Ruhr could, in theory, have been made available, it is not at all certain that, even if strategic developments could have been foreseen in time, the supply apparatus could have adapted itself with sufficient speed or displayed the necessary determination. Given the excessive conservatism and even pusillanimity that characterized the logistic planning for 'Overlord' from beginning to end, there is good reason to believe that this would not have been the case. It is impossible to imagine the prudent accountants, who considered the advance to the Seine impracticable even while it was being carried out, suddenly declaring themselves willing to take the risk of supporting an 'unscheduled' operation across the German border. That the SHAEF logisticians were not cast in the heroic mould it seems impossible to deny. Yet it is hardly for us, who have seen so many great campaigns come to grief owing to a lack of logistic support, to condemn one which did after all terminate in an undisputed – if possibly belated – success.

8

Logistics in perspective

Looking back over the present study, particularly its latter chapters, it sometimes appears that the logistic aspect of war is nothing but an endless series of difficulties succeeding each other. Problems constantly appear, grow, merge, are handed forward and backward, are solved and dissolved only to reappear in a different guise. In face of this kaleidoscopic array of obstacles that a serious study of logistics brings to light, one sometimes wonders how armies managed to move at all, how campaigns were waged, and victories occasionally won.

That all warfare consists of an endless series of difficulties, things that go wrong, is a commonplace, and is precisely what Clausewitz meant when talking about the 'friction' of war. It is therefore surprising that the vast majority of books on military history manage to pay lip service to this concept and yet avoid making a serious study of it. Hundreds of books on strategy and tactics have been written for every one on logistics, and even the relatively few authors who have bothered to investigate this admittedly unexciting aspect of war have usually done so on the basis of a few preconceived ideas rather than on a careful examination of the evidence. This lack of regard is in spite – or perhaps because – of the fact that logistics make up as much as nine tenths of the business of war, and that the mathematical problems involved in calculating the movements and supply of armies are, to quote Napoleon, not unworthy of a Leibnitz or a Newton. As a great modern soldier has said:[1]

> The more I see of war, the more I realize how it all
> depends on administration and transportation. . .It takes
> little skill or imagination to see *where* you would like your

army to be and *when*; it takes much knowledge and hard
work to know where you can place your forces and whether
you can maintain them there. A real knowledge of supply
and movement factors must be the basis of every leader's
plan; only then can he know how and when to take risks
with those factors, and battles are won only by taking risks.

The history of logistics has been divided into periods according
to two main criteria. Following the tradition originating in
Clausewitz and Moltke, some writers have identified three distinct
periods of logistic history in modern times, based on the supply
systems used. The first of these comprises the age of standing
armies, when military forces were magazine-fed, the second
embraces Napoleonic 'predatory' warfare, and the third, opening
in 1870–1, is marked by a system of continuous supply from base.
Variations on this classification have occasionally been produced,
e.g. by one authority who regards the *Etappen* system as 'essentially a modification of the rolling magazine of the eighteenth
century'.[2] This implies that the development of logistics was,
apart from a short and regressive period around the beginning of
the nineteenth century, a smooth and continuous process.

Alternatively, inquiring into the causes behind the successive
developments in logistics, other authors have concentrated on the
technical means of transportation employed. This made it possible
to divide the history of warfare into neatly compartmentalized
periods: the age of the horse-drawn wagon was succeeded by that
of the railway, which in turn was superseded by that of the motor
truck. Though each of these methods of transportation had its
own characteristic qualities and limitations, the overall trend was
nevertheless towards carrying ever greater loads at ever higher
speeds. With the exception of Napoleonic practice, the development of logistics could be pictured as a continuous process.

In the light of our investigation of the details of the logistics
systems employed during the last century and a half, there does
not appear to be much accuracy in either of these classifications.
The image of eighteenth-century warfare as magazine-fed, its
movements shackled by its own supply apparatus, is largely false.
Napoleon, far from regressing to more primitive methods, in fact
made more extensive use of his predecessors' normal practice and
ended up by constructing what was by far the most comprehensive system of supply the world had ever seen. *Pace* Moltke and

the demigods, the arrival of the railways and the *Etappen* system did not effect any revolution in the supply of mobile operations, with the result that, even as late as 1914, the German troops were expected to live off the country and actually did so. Seen as an exercise in more or less well-organized plunder, the history of warfare from Wallenstein to Schlieffen forms an easily-recognizable whole. The period is characterized above all by the fact that armies could only be fed as long as they kept moving. Once they came to a halt in front of either Mons (1692) or Metz (1870), however, they experienced immense difficulties, which had to be overcome by resorting to 'artificial' means. This was so even in 1914, as German troops investing Antwerp found to their cost.[3]

If this continuity was finally broken in 1914, this was due not so much to any sudden humanization of war, as to the fact that, following the enormous rise in the consumption of ammunition and other prerequisites of war (including, for the first time, motor fuel), armies found themselves no longer able to take the majority of their supplies away from the country. Whereas, even as late as 1870, ammunition had formed less than 1 percent of all supplies (6,000 tons were expended as against 792,000 tons of food and fodder consumed), in the first months of World War I the proportion of ammunition to other supplies was reversed, and by the end of World War II subsistence accounted for only eight to twelve per cent of all supplies. These new demands could only be met by continuous replenishment from base. Thus it now became relatively easy to support an army while it was standing still, almost impossible to do so when it was moving forward fast. It is not surprising that it took time for this reversal of all previous notions to be understood.

Until the problems thus generated could be overcome, a type of warfare evolved which, precisely because the railways could *not* replace the horsed wagon, entailed supplies and sometimes troops, being kept in the same place for literally years on end. Huge quantities of supplies (e.g. one-and-a-half million shells for the single offensive launched by the British on the Somme in 1916) were brought up the front, dumped at the railheads, and could never again be moved forward, backward or laterally. To a far greater extent than in the eighteenth century, strategy became an appendix of logistics. The products of the machine—shells, bullets, fuel, sophisticated engineering materials—had finally superseded those of the field as the main items consumed by

armies, with the result that warfare, this time shackled by immense networks of tangled umbilical cords, froze and turned into
a process of mutual slaughter on a scale so vast as to stagger the
imagination.

Whether or not the subsequent products of the machine –
trucks, tracks, airplanes – have enabled armies to overcome the
effects of mechanized warfare is a moot point. Comparisons
between modern mobile campaigns and those of earlier days
have sometimes been made, but their results are hardly convinc-
ing in view of the difficulty of finding the right things to compare.[4] A more fruitful approach to the problem would seem to be,
not to measure the speed of present-day armies as against that of
previous ones, but to try and see to what extent they have been
able to realize their maximum theoretical speed as determined by
the technological means available – in other words, to measure
the effects of that elusive factor, 'friction.' Looked at in this way,
it appears that the maximum sustained speed of eighteenth-
century armies was three miles per hour – the pace of a marching
man – and that they were occasionally able to cover as much as
fifteen miles per day for two or three weeks on end. As against
this ratio of 1:5, modern vehicles – even tracked ones – can easily
cover fifteen miles an hour, but no army, not even the British
'Blade Force' chasing Rommel in North Africa or Malinowsky's
armoured columns in pursuit of the Japanese in Manchuria, has
managed to sustain a pace of seventy-five miles a day for more
than a very few days at a time.

Another way of looking at the same problem is to consider the
so-called 'critical distance', i.e. the maximum one at which armies
can, with the aid of a given type of vehicle, be 'effectively' supported from base. We have already had cause to remark that the
concept itself is not very useful, dependent as it is on a very large
number of accidental factors. Nevertheless, a theoretical examination based on it appears interesting. Assuming a wagon of one-ton
capacity drawn by four horses, each of which consumes 20lb. of
fodder per day, the maximum distance the vehicle can travel
before using up its entire payload is $\frac{20 \times 2,240}{80}$ =555 miles, of
which probably no more than 120, or approximately twenty-two
per cent, have ever been utilized in practice. Compared with this,
a 5-ton World War II motor truck loaded with nothing but fuel
could probably travel at least 5,000 miles before getting through

its load. However, no more than, at the most, 500 – or ten per cent of the maximum capacity – have ever been utilized in a European campaign. Even Rommel, a commander who understood the art of squeezing the last mile out of his transport like no other, ran into insoluble difficulties when he tried to double this figure. It follows from these facts that, far from eighteenth-century armies being slow and magazine-bound, they managed to do very much better in relation to the theoretical limits of the means at their disposal than do modern forces. This is only to be expected, for the friction within any machine – human or mechanical– increases in proportion to the number of its parts – a prime example of the law of diminishing returns.

Given these facts, and in spite of the development of all means of transportation in the thirty years since World War II, the speed of mobile operations cannot, in our opinion, be expected to rise dramatically in the foreseeable future. It was demonstrated in the Israeli–Egyptian campaign of 1967 that not even an army operating under ideal conditions, enjoying complete superiority in the air and taking a second rate opponent by surprise, stands much chance of covering more than forty miles per day.[5] To conclude from this that the 1944-style army is 'dangerously obsolete', however, is to misunderstand the development of logistics during the present century, characterized as it is by the fact that the speed and range of successive new means of transport have been largely, if not completely, offset by the enormous increase in friction and, above all, by the quantities of supplies required. Hence, even if an entirely new and unprecedentedly effective means of transport were to appear tomorrow, it is to be expected that only a small fraction of its maximum theoretical capacity will ever be utilized in practice, and that its effect on the speed of mobile operations will therefore be marginal.

A further interesting question concerns the proportion of service to combat formations. This proportion is frequently cited as a rough indicator of an army's efficiency – a low proportion representing a high efficiency. But this is to misunderstand the relation of service to combat units. Romantically heroic politicians and gung-ho generals notwithstanding, the aim of a military organization is not to make do with the smallest number of supporting troops but to produce the greatest possible fighting power. If, for any given campaign, this aim can only be achieved by having a hundred men pump fuel, drive trucks and construct railways behind each combatant, then 100:1 is the optimum ratio. To work out this ratio with its thousands of component parts in theory, however, is an almost impossible task, involving as it does a model of tremendous complexity, worthy of a battery of computers presided over by a mathematical prodigy. Moreover, when all the calculations have been done, and the preparations made on the basis of them completed, there is always a strong possibility that a fresh strategic or political requirement will render them worthless.

In practice, there is scant evidence that the task has been attempted by the majority of twentieth-century (not to mention earlier) operational planners. Rather, most armies seem to have prepared their campaigns as best they could on an *ad hoc* basis, making great, if uncoordinated, efforts to gather together the largest possible number of tactical vehicles, trucks of all descriptions, railway troops, etc., while giving little, if any, thought to the 'ideal' combination which, in theory, would have carried them the furthest. Nor, as we saw, were the results of the only comprehensive attempt that *was* made in this direction particularly encouraging. In spite, or perhaps because, of the fact that the plans for 'Overlord' made detailed provisions for the last prepacked unit of fuel, they quickly turned out to be an exercise in conservatism, even pusillanimity, such as has not often been equalled. Not only did the actual development of the campaign have little in common with the plans, but the logistic instrument itself functioned very differently from what had been expected. Consequently, it would hardly be an exaggeration to say that the victories the Allies won in 1944 were due as much to their disregard for the preconceived logistic plans as to their implementation. In the final account, it was the willingness – or lack of it – to override the plans, to improvise and take risks, that determined the outcome.

It is perhaps fitting that the present study – starting, as it did, with the determination to avoid 'vague speculations' and concentrate on 'concrete figures and calculations' – should end with an admission that the human intellect alone is not, after all, the best instrument for waging war and, therefore, understanding it. The intellect must play a central role both in planning and execution, if only because no better instrument is available to us; but to believe that war, or indeed any other aspect of human behaviour, can ever be grasped by means of the intellect alone is to give proof of a *hubris* like that evinced by those who built the Tower of Babel and deserving similar punishment. To recognize the truth of Napoleon's dictum that, in war, the moral is to the physical as three to one; in the final account, this may be all that a study of the influence of logistics on mobile operations may be able to teach us.

POSTSCRIPT: WHERE ARE WE NOW?

Almost thirty years have passed since, living in London and working mainly at an old dressing table of my landlady who has long since passed away, I wrote *Supplying War*. Thirty years is a long time; while some things have remained more or less the same, others have changed beyond recognition. Against this background, the present postscript attempts to do three things. Part I asks what has happened to the history of military logistics as a field of study. Part II asks where *Supplying War* itself stands amidst the rapidly growing literature, including some that is critical of it. Part III takes a brief look at the post-1945 development of military logistics themselves. What has happened to them, and where are they headed?

I

To start, then, with what happened to logistics as a field of study. As even a cursory look at the catalogues will show, the most obvious change has been the amount of attention paid to the issue. Thirty years ago the literature on the history of military logistics was extremely limited; indeed almost the only group attracted by the subject were a few Austrian-Hungarian officers who, for some obscure reason, had done a considerable amount of work on it between about 1866 and 1914.[1] Then as now, discussions of warfare tended to revolve around the marvels of ever more sophisticated weapons and weapon systems, a term that was just coming into vogue. Others wrote of armies that advanced and retreated, maneuvered and outflanked, encircled and penetrated; by contrast, logistics were not considered sexy. In part, this may have reflected the impact of the most influential twentieth-century military pundit, Basil Liddell Hart, whose work on strategy first saw the light of print in 1929 and continued to be published each time a major conventional war broke out anywhere in the world. Unfortunately he had died some years before I came to England so I did not have the privilege of meeting him. However, his legacy was intact and his light was marching on.

Be this as it may, very few military historians at the time asked themselves what armies on campaign required, how they got what they required and how those requirements affected the things they did or did not do. Even fewer tried to answer these questions in ways that were more than superficial. Take David Chandler's *The Art of Warfare in the Age of Marlborough* (1976). Nobody knew more about eighteenth-century warfare than David, whose book is superb in many ways, yet anybody who searches the index of this book for the term *logistics* will do so in vain. I also remember my disappointment after reading Field Marshal Bernard Montgomery's *Concise History of Warfare* (1972) while I was working on the logistics of the Schlieffen Plan. Montgomery himself, I knew, was a thorough professional who left as little to chance as he could. Although in his memoirs he pretends to be above such things as the precise way his armored cars received their petrol,[2] he certainly did not neglect supply; had he done so, he would never have marched from Alamein to Bizerta or landed on the Normandy coast. All the greater, therefore, my chagrin at finding that he had almost nothing to say about the subject. What he did say was based on a conversation with Liddell Hart, who did not have much to say either; focusing on strategy, the master treated war almost as if it were a game of football which does not include logistics. At the time, personalities still counted for something and I, a young scholar, was too polite to take them to task. *Faute de mieux* I was compelled to cull my material from numerous unrelated, often obscure, sources. I spent months pouring over musty tomes at the British Museum and other London libraries; what was one to make of the letters of some seventeenth-century *Marechal de France* whose very name hardly anybody recalled? At other times I sat in the reading room of the Bundesarchiv/Militaerarchiv in Freiburg, desperately contemplating mountains of Wehrmacht documents and trying to decide which to read and which to put aside. To those who helped me with that and other tasks, I am still grateful.

Since then, what a change! The literature is now vast beyond measure[3] and I flatter myself that some small part of it owes its inspiration to my work. Here and there authors have turned to me for help, which I have been glad to give. Partly to make room for the countless people who now study it, partly as a result of the usual scholarly quest for order and comprehensiveness, the field has been widened far beyond Jomini's definition of logistics I used. It now includes naval warfare, air warfare, planning, war production, procurement, administration and a host of other subjects.[4] Some authors include finance, which has been called the sinews of war, and which in turn causes them to look at taxation and other methods of raising money. Others delve into specialized fields such as the construction of roads and bridges, research and development, and what not; it has even been claimed that, since constructing fortifications clearly requires a considerable logistic effort, works on the way war was supplied should include fortification.[5] Some of the research borders on the trivial. It is true that weight alone does not adequately express the relative importance of some kinds of supplies. Some campaigns have been lost for want of a nail; on the other hand, when one-time Soviet dictator Nikita Khruschev asked whether the fact that soldiers cannot fight with their trousers down should cause buttons to be declared strategic goods he had a point. Like economics, the dismal science in which it is rooted, logistics, and even more so the study of logistics, has the ability to expand until it swallows almost everything else. Much of this expansion is natural, even necessary. Still, in some cases it goes too far. Does one have to remind readers of Clausewitz's dictum that, at the very heart of war, there is fighting?

As an example of the way the literature has exploded, take a subject I did not explore, i.e., the logistics of the Roman Army. At the time I wrote, the only available work was Anton Labisch's *Frumentum Commeatesque* (1975), a pioneering volume which came to my attention too late to use. Since then our knowledge and understanding of the subject has been immensely enhanced. First came Paul Adams's *Logistics of the Roman Imperial Army* (1976). Next, to list the principal works only, there appeared James Anderson's *Roman Military Supply in North East England* (1992); Theodor Kissel's *Untersuchungen zur Logistik des roemischen Heeres* (1995); Jose Remesal-Rodriguez's *La annona militaris* (1986); Paul Erdkampf's *Hunger and the Sword: Warfare and Food Supply in Roman Republican Wars* (1998); and finally Jonathan Roth's comprehensive *The Logistics of the Roman Army at War* (1999). Not to mention any number of articles in scholarly journals on subjects from the carrying capacity of ancient pack animals to the methods the Legions used to requisition food.

Though this has little to do with military history, another notable development of the last thirty years has been the emergence of logistics as a separate, and quite important, academic field. Whether in a military context or a civilian one, it is being studied by tens of thousands of people all over the world. Scarcely a week passes without another center, or department, or institute opening its doors and calling on those with money in their pockets to enter and look at the marvels on display. Indeed the term 'logistics' itself has acquired a sophisticated ring that is part of the subject's attraction. Tell people you are a 'logistician' – so much better than a mere warehouseman, or porter, or driver, or bean counter – and they will think you are very clever if, perhaps, on the boring side. In German one can even speak of 'the logistics of perception' (*Logistik der Wahrnemung*); what that may mean I have not the slightest idea. Seriously, there are now so many publications each year as to preclude one person surveying, let alone reading, all of them. Nor does the fact that many of them appear on the Internet always make things easier. Internet publications tend to be readily available but inconvenient to use. To use a logistic analogy, it is like providing troops with biscuits instead of bread: although biscuits are cheap and easy to keep, they are hard to digest.

To the extent that it has added enormously to our knowledge, this mighty flood of publications is, of course, welcome. There is, however, another side to the coin. When I first entered the field I did so as a complete innocent; the most complex logistic operation I had ever engaged in was to move my family from one country to another. Academically speaking, my background consisted of one or two long-forgotten books I had stumbled on which made me want to know more about the way armies used to supply themselves and the impact that their doing so has on operations. Since then, the literature has made me realize how utterly impossible it is for anybody to master this vast subject in its entirety. Sometimes I feel like the traveler who, having lost his way in a blizzard in a moonless night, unwittingly walked across a frozen lake. Looking back and realizing what he had done, his heart stopped beating and he dropped dead.

II

Second, the book. Provided the author defines his goals clearly and sticks to them consistently, one must not demand of a book that it do more than it set out to do; to criticize in retrospect is always easy and in some cases doing so is cheap. At the time I produced *Supplying War* I was venturing into an unexplored field and yet attempting to cover a period of more than three centuries; if anything I thought I was being very ambitious. It is true that I left out numerous subjects including naval warfare (as, incidentally, Sun Tzu and Clausewitz also did) and air warfare. It is also true that, out of the three levels – the strategic, operational and tactical – I really only concerned myself with the second. Finally, I gave proof of my intellectual limitations by not studying the logistics of extra-European campaigns including, above all, those waged by the almighty Americans against each other. About all this I can only say, *mea culpa*.

Nevertheless, and even though the amount of detailed knowledge that has been added since then is enormous, I would argue that some of the basic premises which underlie *Supplying War* remain intact. If anything, they have become so self-evident that few people even bother to mention them any longer. In the first place, everybody now knows that logistics are extremely important – or perhaps I should say that, since commanders have always been aware of his fact, this is something even military historians now realize. From time immemorial questions of supply have gone far to govern the geography of military operations, their timing, the things they could do and their outcome; the days when, submitting a paper to a scholarly periodical, I had to apologize for writing about such a marginal subject are definitely gone. Second, and following from the first, everybody agrees that a study of military history that does not take logistics into account is amateurish. This, of course, is a cardinal reason behind the vast explosion that has taken place in the relevant literature.

At a lower level, too, the basic theses of *Supplying War* still stand, and I would like to take this opportunity to amplify them with the aid of some additional case material. First, during most of history, transport by water was much cheaper and easier than by land. Referring to Rome, some authors have estimated the difference at 1:50;[6] later, since ships became larger whereas horses remained more or less the same size, the figure must have grown. In other words, moving a load by pack animal or wagon cost up to 50 times more than doing the same by ship, and the bulkier and less expensive the goods the greater the difference. As a result, most commanders did the sensible thing and preferred to move supplies by water if they could. What land-bound lines of communication existed tended to be short. Take Jonathan Roth, who has done as thorough a study of Roman military logistics as may be had. Yet in the course of about five centuries he studied, he found only three cases when supplies were brought up overland from a base 100 miles behind the lines.[7] Just how it was done cannot be learnt from the sources; the interesting thing is that 100 miles is quite close to the limit I gave for armies that depend on horse-drawn transport. In the vast majority of cases the distances covered were far smaller. As Roth himself says,[8] during Imperial times every legionary base known to us but one was based on a river, and for very good reason.

Second, it remains true that, to quote a famous late-nineteenth-century German officer and military writer,[9] until about 1900 the weight of factory-produced supplies – ammunition, spare parts and the like – was 'as nothing' compared to that of the food and fodder armies needed.[10] By way of an example of the way things worked, consider the Battle of Borodino fought by Napoleon against Tsar Alexander I on 7 September 1812. The number of French troops present was around 133,000. Between cavalry, artillery and trains some 40,000 horses accompanied the army. At a minimum, the daily ration of each soldier weighs 2.6 lb., that of a horse 26 lb. (both figures exclude water); total weight of food and fodder, 692 short tons. On that same day the French army fired 1.2 million musket balls and 60,000 cannon balls. It was perhaps the heaviest firepower brought to bear by any army in any single battle in history until then; the Union troops at Gettysburg only fired half as many balls in three days.[11] At 0.06 lb. per infantry round and 10 lb. per artillery round (both including powder) the total weight came to 336 tons; of this, almost 90 percent was taken up by cannon and just over 10 percent by small arms. Thus, even on that day, far more food and fodder were consumed than ammunition fired. Yet Borodino was the only full-scale battle in an enormous campaign which, starting out with 600,000 troops and 250,000 horses, lasted six months from the crossing of the Niemen to when the last frozen stragglers staggered back over the Russian-Polish border. Anyone who, using these figures, wishes to work out the relationship between the two kinds of supplies during the entire campaign is welcome to do so.

Third, and precisely because the most important items (by weight) could be had wherever there was a dense civilian population, armies engaged in field warfare tended to live off the countryside they traversed. Commanders would have to be foolish indeed to try and bring up from base what could be found on the spot. To return to the above-mentioned campaign of 1812, does anybody seriously believe Napoleon brought up the tens of thousands of tons of food and fodder his army consumed from his great depot at Strasbourg? Or even from the forward base at Koenigsberg? Of course he did not. At most, this applied to the ammunition, clothing, medical supplies, and similar supplies the *Grande Armee* needed and which were stored on the frontier before the campaign got under way; even for these commodities, though, Napoleon relied on captured Russian stores wherever and whenever possible. Though doing so was easier in the relatively densely inhabited regions around Smolensk than it had been further west in the forests of Belorussia, almost everything else was obtained on the way. This is even more true when, supplementing *Supplying War*, one considers firewood as well. From Republican Rome to the nineteenth-century Russian Army, normally there was one campfire (as well as one tent, one cooking pot and one pack animal) for every eight to ten soldiers. Yet I doubt whether, in the whole of history, there was even a single case when an army brought up firewood from the rear as modern armies transport fuel; forests, or, at a pinch, the inhabitants' houses, would do.

The methods by which supplies were extracted differed. Partly to prevent the enemy from interfering, partly to prevent desertion and partly to make sure no supplies were destroyed, pilfered or wasted, a considerable amount of organization was required. One method, especially useful in the case of fodder, was to send out units of troops to harvest supplies from the fields while others remained in formation and stood guard. Another was requisitioning, i.e., ordering the local authorities to deliver so many pounds or units of such and such a commodity at such and such a time and place; this, incidentally, remained the preferred German/Prussian way of doing things even as late as 1870–1.[12] A third was purchase, which might be either free or forced. In the former case the army would set up a market; the latter had much in common with requisitioning. The necessary sums might come from the army's own coffers. More likely, though, commanders would make the local population pay a 'contribution'. They raised money – not worthless pieces of paper, but such as made a clinking, clanking sound – and used it to pay merchants or sutlers. The latter were held responsible for purchasing the supplies and transporting them.

However it was done, an army passing through a province would leave it impoverished if indeed it did not also destroy property and kill whoever refused to surrender it as fast as was demanded of him. All too often this was true even when the province in question was friendly or neutral. Few if any armies campaigned on capitalist principles as the modern American one does, freely negotiating contracts and paying market prices for items delivered.[13] Even the Americans themselves, before they became the richest nation on earth, were not always as conscientious in these matters as they claim to be today. Archives dating to the War of the Revolution are stuffed with thousands of notes promising payment for supplies requisitioned; many are covered by semi-literate scrawls.[14] What took place in enemy territory could defy description. The more lax the discipline and the greater freedom the troops were given to help themselves, the worse the consequences for the local population.

For example, during the Thirty Years War every army had a special official known in German as *Brandstaetter* ('conflagrator'). Accompanied by a guard, he went around captured towns and assessed their value, threatening to burn them down if his demands were not met. Some towns and provinces, hoping to keep the worst evils away, offered the armies that approached them 'voluntary' contributions; as one would expect, some of those contributions later became regular and obligatory. From at least Roman times on there were also cases when local authorities tried to bribe commanders and supply officials to make sure the lands under their jurisdiction would be spared. One early eighteenth-century German Countess wrote to the Duke of Marlborough, asking him to steer his army away from her estate. He and his officers, of course, were welcome to visit; it was true that the Count was absent, but he would understand.

The fact that armies depended on local supplies did not mean that they could not be deprived of them. Foraging parties could be ambushed, either as they dispersed to do their work – of which it is possible to find many examples in ancient warfare – or else on their way back to camp when they were heavily laden. The enemy might also try to cut communication between an army and the place where it had gathered, or ordered, or bought its supplies; in Latin, the *terminus technicus* for this is *intercludere*. Some commanders deliberately maneuvered their enemies into districts where supplies were unavailable, leaving them with the choice of breaking out or surrendering. On other occasions, provinces were 'devastated' to prevent the enemy from using the supplies they contained. Examples of the last-named method may be found all the way from Fabius Maximus in 217 B.C.[15] to the nineteenth century and beyond. Peter the Great used it to delay the invasion by Charles XII, leading to the comment that he was acting like a man who cut off his nose to spite his wife.

On the other hand, and barring special conditions such as forests, mountains, and, above all, deserts, the ability to gather supplies locally gave armies a degree of freedom that their modern successors can rarely match; often the solution to supply difficulties in one district was simply to move into another. As one expert wrote of ancient Greece,[16] warfare consisted of extended walking tours accompanied by large-scale robbery. Donald Engels's excellent volume on the logistics of Alexander the Great[17] is the most successful attempt to show how it was, or, since our sources are not really sufficiently detailed, could have been, done. The King and his army embarked on an unprecedented adventure into semi-legendary lands. As they prepared to set out, nothing was more important than acquiring advance knowledge of where food, fodder, and, above all, water could be found. This information in turn permitted arrangements to be made with local officials who, no doubt after being thoroughly intimidated, delivered supplies along the road ahead. Money, which kept falling into Alexander's hands as he captured Darius's treasure chests, helped. Had it not been for this system the army would have starved to death; this very nearly happened during the march through the Gedrosian Desert where there were no towns or villages and where neither money nor threats could induce supplies to come forward. The ability of armies to navigate inhabited territories almost as navies do at sea is also demonstrated by the fact that, within the period covered by *Supplying War*, Gustavus Adolphus and the Duke of Marlborough, and, much earlier, Edward III of England and Alexander fought battles with 'reversed' fronts. Had lines of supply played as great a role as they did later, doing so would have been impossible. Instead we would have spoken of the commanders in question as facing a desperate situation; and they themselves, aware of that situation, might have surrendered rather than stood, fought and conquered.

More significant still, these and many other commanders sometimes marched so far away from their home countries as to almost lose touch with them. Their only means of communication consisted of letters that took weeks to arrive. Even when they did arrive, the most they got from the homeland were occasional reinforcements and, perhaps, sums of money. Logistically as well as in other respects they were almost completely independent. They extracted supplies from the country they had traversed or occupied, built up their own bases on the spot, and, depending on how rich the country was and how successful their own methods, had their troops live either in plenty or in penury. The logistic independence they enjoyed also meant that if they were not already crowned when they set out, they were sometimes able to return as would-be military dictators;

as, limiting ourselves to Rome, Marius, Sulla, Pompey, Julius Caesar, Vespasian, and any number of third-century emperors did. Clearly logistics can only go so far in explaining such events, which were occasioned by many other factors as well. On the other hand, none of it would have been possible if the commanders in question, instead of building up their own bases from local resources, had been tied to regular lines of supply leading from the home country to where they were.

Not only was field warfare mobile by definition, but some of the mobility was specifically occasioned by the need to keep the army supplied. Indeed the Latin term for military supply (*commeatus*) can also mean 'the ability to move about freely'. It is as if the language itself tried to tell us that the former could only be had by means of the latter. When Livy, or Caesar, or Tacitus, used the word, it may be hard to say which of the two meanings is intended; yet Latin, not surprisingly, is notorious for the number of military terms it contains and the many nuances those terms are able to express. By contrast, siege operations might involve staying in the same district for weeks, months, or, though this was exceptional, even years. To make things worse, the district in question would often have been denuded of supplies ahead of time by the besieged as part of their own preparations. The fact that siege operations were continuous, and the drain put on local supplies, made armies that engaged in them much harder to maintain than those that campaigned in the field. Already in ancient Rome the sources often refer to the fact that every siege was really a race between the besiegers and the besieged to see who would starve first.[18] No wonder that, from the second half of the seventeenth century on, some of the earliest known modern logistic organizations were put together specifically to besiege cities.

Fourth, and as the discussion of Borodino already implied, historians have tended to overestimate the extent to which French Revolutionary and Napoleonic warfare developed a new logistic system. As far as requirements went, the logistics of the *Grande Armee* were no different from those of its predecessors, men remaining men and horses horses. Nor, when it came to meeting those requirements, did the French enjoy any important advantage; if anything, Napoleon tended to be conservative, technologically speaking. Like almost all his predecessors the Emperor did what he could to make war feed war, employing a specialized organization whose task was to commandeer supplies and money wherever possible. Where he differed from most was that his armies were, thanks to conscription, considerably larger. Size alone dictated that the *Grande Armee* be divided into large units – known as *corps* – and that those units should march not along a single road but along several different ones. Marching along several different roads made movement easier, explaining some of the speed for which the French were famous. More important in our context, spreading out made it much easier to find supplies. As I have shown elsewhere,[19] Napoleon's real innovation consisted not of his logistics but of his command and control system. The latter kept the entire enormous machine together, enabling him to coordinate his formations in spite of the relatively vast distances that separated them from each other and from central headquarters. Finally, spreading out also had the advantage of enabling the army to bypass most fortresses. Thus avoiding precisely those siege operations which, traditionally, had imposed the heaviest logistic burden.

Fifth, the introduction of the railways initially did less to affect army logistics than many historians have thought. By making possible the modern system of conscription and trained reserves, railways revolutionized the way armies were raised and organized. Equally revolutionary was the way they were deployed on the frontiers of the countries to which they belonged; in 1866, which was when the Elder Moltke showed how it should be done, the rest of the world held its breath. Railways also enabled entire armies to be shifted from one theater of operations to another, a fact without which the American Civil War in particular would have taken a very different form. Finally, railways made their own logistic demands in the form of wood or coal, water, spare parts and the like. Relative to the loads that they could carry, though, those demands were far smaller than those of the horse-drawn wagons they replaced or the motor vehicles that eventually replaced them. The difficulty with railways was that, once active campaigning got under way, they had trouble keeping up. Reconstructing a damaged line, let alone building a new one from scratch, was usually a prolonged and complicated affair. Until this was done armies had to manage as best they could. Because their most important requirement (by weight) still consisted of food and fodder, this was something they were usually able to do.

What were the logistic requirements of a mid-nineteenth-century army operating in the field? To answer this question at the hand of a conflict *Supplying War* did not consider, take the Seven Days' Battle (June 1862), which saw the participation of almost 100,000 Federal troops accompanied by no fewer than 40,000 animals. Daily consumption of all types of supplies is said to have stood at 600 tons.[20] Assume the same 26 pounds per animal and 2.6 per man that we calculated for Borodino; in so far as Grant fed his troops much better than Napoleon did,[21] the latter figure is almost certainly an underestimate. Multiply 26 by 40,000 and 2.6 by 100,000 and add the results. The answer is 650 (short) tons. While the sources do not provide sufficient data to explain the discrepancy, clearly even at this late date food and fodder together overshadowed everything else to leave room for ammunition as well as many other kinds of supplies; some, perhaps much, of the fodder was not carried by the army's trains but had to be gathered on the spot. Since fodder is more readily available in some places than in others, clearly this issue must have had some impact on where armies could and could not go, how long they could stay and what they could or could not accomplish while they did stay. Finally, to put the matter into perspective, during the entire four-year Civil War there was only a single Seven Days' Battle. Most days passed with no major fighting at all; what this must have meant for logistic demands and the ability to meet them also seems quite clear.

Throughout the two millennia in question, the single most important item – in terms of bulk – was usually fodder. While not all types of fodder could be taken directly away from the fields, many could be and were. Reflecting these basic facts, the term 'to forage' came to stand for the local acquisition of supplies of any kind; this was as true in German (*fouragieren*) and English as in the original French. A perfect example is provided by the Union Commander General William Sherman. Preparing for his epic march through Georgia to the Atlantic Ocean in January 1864, he wrote: 'each brigade commander will organize a good and sufficient foraging party . . . [It] will gather near the route traveled, corn or forage of any kind, meat of any kind, vegetables, corn meal . . . to keep in the wagons at least ten days provisions.'[22]

As *Supplying War* explains in some detail, the conflict that really caused all this to change and ushered in the modern age was World War I. The period from 1870 to 1914 saw tremendous advances in military technology, particularly the rise of the magazine rifle, machine guns, and, above all, quick-firing artillery. Though railways also made progress in terms of both numbers and carrying capacity, they were unable to keep up with the new logistical demands that these machines made or the massive increase in the size of armies. Along with the superiority of the tactical defense, on which much has been written, and the difficulty of adapting wire-bound (telegraphic) communications to offensive warfare, about which I myself have written,[23] these factors played a major role in the rise of trench warfare. Trench warfare by definition meant staying in the same place for an extended period, making it hard to rely on local supplies; the more so because the areas immediately behind the front would have been evacuated and devastated. In addition to ammunition it made vast demands in terms of barbed wire, construction materials and the like. All these had to be produced in the rear, even if the rear itself sometimes consisted of occupied territory. Next they had to be loaded on trains, transported and unloaded at the railheads which, for the sake of security, tended to be located beyond artillery range. While motor vehicles were already in use, their number was limited and they only played a comparatively small role. Hence the last leg had to be covered by the time-honored means of horse-drawn wagons and carts, sometimes supplemented by light field railways which could rarely enter within the range of field artillery. To make things worse still, each major offensive started with the firing of hundreds of thousands, sometimes millions, of shells. By churning up the roads and turning the country into a moonscape of craters, they made it even harder to carry out a successful advance.

Finally, the large-scale use of motor vehicles during World War II eased the constraints imposed on operations during World War I. Along with improvements in command and control, this ushered in the era of armored warfare and the famous *Blitz* campaigns, when tens of thousands of tanks fought each other from Normandy to Stalingrad and from Alamein to Milan and beyond. Logistics did, however, impose limits on the operational freedom even of the best-commanded, best-equipped armies. In part, this was because almost all the supplies an army required – around 90 percent – now consisted of factory-produced items which could only be procured far in the rear and, once this had been done, had to be transported to the front. In part it was because the principal means that was used for this purpose, i.e., motor vehicles, made their own not inconsiderable demands in terms of manpower, POL (petrol, oil, lubricants) and spare parts, demands which once again had to be answered from base.

Between them, those developments led to two opposing results. On the one hand they meant that armies found it easier to operate in environments, such as deserts, that had previously been inaccessible to large units; imagine the Afrika Korps without motor transport. On the other hand, the days when logistic problems could often be solved by marching the army from a province that had been eaten bare to another that still contained provisions did not return. The freedom of movement armies had enjoyed until the end of the nineteenth century was not restored. Perhaps more than during any previous period, field armies were tied to 'umbilical cords of supply'. Losing those cords meant swift disaster. Ask the 250,000 German troops who, having starved for weeks, were either killed or taken prisoner at Stalingrad. Or, for that matter, the troops of the U.S Third Army who had to halt their advance into Lorraine not because the Wehrmacht kept fighting but because their tanks ran out of fuel.

When *Supplying War* was written, and certainly when I attended the university as a young student during the 1960s, academic historians tended to be a more cautious, less pugnacious lot than they are today. The prevailing wisdom was that history had ended in 1914, or 1945, or whenever. Many thought that the term 'contemporary history' was a contradiction in terms. Those who wrote about it were really journalists, who, in turn were little better than hacks. Having already stuck my head into the lion's mouth by entering a largely unexplored field, I did not dare venture further into the present than 1944. In any case I suspected that, if I *had* done so, I would have found little that was fundamentally new; studies of three armored campaigns, i.e. those that took place in Russia in 1941, in North Africa in 1941–2 and in France in 1944, seemed quite enough. All this explains why the volume ended with World War II, and not with Korea, or the June 1967 Arab-Israeli War, or even the October 1973 one. Once again, I can only say *mea culpa*.

III

How, then, did military logistics develop from 1945 on? Though this was by no means immediately obvious, and though many people continued to think in terms of 'total' war,[24] looking back we can see that World War II was a historical turning point. Until then conventional interstate war had been growing for centuries. Though there were some ups and downs, in the end it always involved larger and larger armies. The armies demanded more and more supplies to support the larger and larger campaigns on which they engaged; culminating with Hitler invading Russia with three and a half million men. As the most powerful states followed each other in building nuclear weapons, however, wars of this kind became few and far between.

Those that did occur tended to be fought against, or between, third-
and even fourth-rate military powers. Consider the Arab-Israeli wars,
or, an even better example, those waged in the Persian Gulf first by
Iraq against Iran and then by the United States against Iraq. Though
the global population tripled between 1945 and 2000, the number of
men and women wearing military uniforms tended to decline, a process that affected first so-called developed Western countries, then
former East Block ones and finally developing ones such as China
and India.[25] As the principal armed forces shrunk, in some cases to
less than 5 percent of their former size, the amount of supplies they
needed to keep their personnel fed and their machines running also
decreased.

The shrinking of conventional interstate war and the logistics it
involved was not a simple process. In part this was because late-
twentieth-century armed forces were much more capital-intensive
than their predecessors. Even as late as 1939–45, World War II was
still fought largely by infantrymen. Except for the Americans, who
had the semi-automatic M-1, most of them still carried ancient bolt
action, single-shot rifles. Moreover, whereas G.I.s and Tommys were
usually carried in lorries, the Ivans, and even more so the Jerrys,
covered enormous distances on foot. I can still see in front of me a
former Wehrmacht officer I used to know. Forty years after the event,
his eyes shined as he told me how, in 1941, he and his comrades had
walked all the way to Moscow. Talk of modern transport as an aid to
strategic mobility; not to mention the fact that, just a few years later,
they walked the entire distance back! Only ten years later, all this
had gone by the board. Not only had all modern armies re-equipped
themselves with automatic rifles, but foot-infantry had almost disappeared. Everybody now rode motor transport and, increasingly,
armored personnel carriers.

Apart from transport, the smaller armies became, the greater,
proportionally speaking, the number of heavy weapons such as tanks
and self-propelled artillery pieces because. Both on the ground and
in the air, the machines themselves tended to become much bigger
and heavier. Take the case of the most prominent machine of all, the
tank. Tanks made their first appearance on the battlefield in 1916.
By the end of World War II most weighed in at around 30 to 35 tons,
though a few, such as the German Tigers and Soviet Stalins, were
heavier. The guns they carried had a caliber of 75 millimeters or so,
and they were propelled by engines capable of producing around 300
horsepower. Over the next 45 years these figures increased to more

than 60 tons, 120 millimeters (which, taking volume and increasing length into account, could mean rounds two to three times as heavy as their predecessors) and as much as 1,200 to 1,500 horsepower. The heavier the machines, the greater the demands they made in terms of fuel, ammunition, and, sometimes, spare parts and maintenance. And, of course, the more of them a unit had the more true this was.

Not only did the logistic requirements of most weapon systems increase, but the entire question became much more complex. Premodern armies only consumed a relatively small number of different items, and indeed much of their supplies came in the form of bulk commodities. Of the items they did require, food and fodder, which as we saw were the most important in terms of weight, could usually be taken from the surrounding countryside by one method or another. Other items could be improvised or manufactured on the spot; every army was accompanied by its own locksmiths and shoemakers and, in the form of female camp followers, persons who could mend clothes. This is seldom the case with modern armies. While they do of course have mobile repair shops, they require vast amounts of different kinds of ammunition and a vast number of different kinds spare parts for the vast number of different vehicles, weapons and other kinds of equipment they use. Practically all of the items in question need to be precision-manufactured. Often this is done to very fine tolerances measured in one ten-thousandths of a millimeter; other items will not survive the slightest speck of dust. After leaving the factory, a great many require specialized storage or have a limited life, after which they become useless or even dangerous. In this way, keeping the forces operational – making sure that the right items will be available to the right unit at the right place at the right time – has developed into a momentous exercise in coordination. In fact, so complex is the problem that it can only be solved, if at all, with the aid of some of the most advanced computers employing the most advanced algorithms, computers which, should the enemy resort to 'information warfare', will themselves constitute vulnerable points.

Thus, although means of transportation have made vast advances since 1945, it does not appear that either the logistic burden has been eased or armed forces have increased their operational freedom. Take the First Gulf War 1991. At the strategic level,[26] the logisticians of Central Command availed themselves of much larger aircraft requiring much larger airports and much larger ships requiring much larger docking facilities. This made the concentration of half a million American troops, plus their supplies, on the other side of the globe much easier and faster than it had been in 1942–4. As the man in charge, Lieutenant General William Pagonis, proudly wrote, in only six months he was able to get more than a million tons of goods into Saudi Arabia.[27] Of course the fact that there were neither German submarines prowling around nor any beaches to storm against Japanese opposition helped; in one sense, instead of waging war, all the Americans were doing was engaging in the kind of large-scale engineering exercise in which they have always excelled. Things were facilitated by the fact that, as was also the case with tanks, the essential characteristics of the various platforms had changed little, if at all. By this I mean that ships remained ships and aircraft remaind aircraft. Their relative capabilities being what they were, more than four-fifths of all supplies required first by Operation Desert Shield and then by Operation Desert Storm were delivered by sea rather than by air. In the future, anyone who is responsible for supplying large-scale military operations overseas will certainly do the same.

Once the supplies arrived at Dahrain and were unloaded, they had to be put aboard fleets of motor trucks and taken to so-called logbases behind the front. Had the war taken place elsewhere, e.g., in Western Europe, then the railways might have played a greater role in the service of logistics. As *Supplying War* points out, railways, though their capacity is much greater than that of motor vehicles, are a rather inflexible instrument. Not only are they less capable of following armies than motor transport, but they are more exposed to enemy action and, in particular, interdiction from the air. Although in northeastern Saudi Arabia, there was no question of the Iraqis striking at Allied logistics in this or almost any other way, on the other hand there were no railways either. Day and night, the convoys, made up of thousands upon thousands of vehicles, roared first across the Saudi deserts and then, after active operations had started, into those of southern Iraq and Kuwait. At peak, one truck was passing a certain point every three seconds; so relentless was the traffic that, to cross the road, people had to use a helicopter.[28] Here and there the trucks were supplemented by transport aircraft and helicopters which brought up supplies by air, resupplying units in need and helping them sustain their advance. However, transport aircraft require fairly extensive ground services, such as runways, control facilities, unloading facilities and the like; they are also too large and vulnerable to venture close to the front. Helicopters can land almost anywhere, but they can only carry fairly small loads at horrendous cost in fuel, spare parts and maintenance. For these reasons, in terms of the tonnage delivered, their contribution did not measure up to the more traditional motor vehicles. In any case, aircraft capable of delivering supplies, and subject to much the same limitations, had already available in World War II; albeit they were much smaller and slower. In these and other ways Operation Desert Storm, for all the vast improvements that have taken place in detail, was merely a repetition of the time when Patton, in his own words, had 'toured France with an Army.'

Another thing that only underwent minimal change was the nature of the supplies themselves. As we saw, in a motorized force such as the American one at the end of World War II, the bulk of the supplies consisted of ammunition, POL and construction materials. By contrast, food (and, even more, so, fodder, the demand for which had all but disappeared) comprised a very small proportion – perhaps 10 percent, perhaps less. The addition after 1945 of more machines, many of which were also much larger, increased the demand for manufactured items. All this caused the century-old trend whereby the importance of the latter increased even as the relative quantity of supplies consumed by people declined to intensify even further. By 1991 POL, much of it consisting of high-quality products such as aviation fuel that could not be locally procured, had become the bulkiest single product by far; only then came ammunition, especially artillery rounds, followed by everything else.[29] Expecting tougher Iraqi resistance and a longer war, Pagonis and his men grossly overestimated requirements and filled the logbases with mountains of everything from hamburgers to artillery rounds and from aviation fuel to aspirins. What made this War unique was the fact that, after it ended, every single item not consumed or expended had to be accounted for, restored to a pristine condition and evacuated.[30] The days when American forces leaving a theater of war would leave behind vast junkyards for the locals to plunder were over, this being itself a symptom of the fact that warfare was becoming both smaller and much more capital-intensive.

While the logbases were bursting at the seams, closer to where the armored spearheads were things did not always look as good. In point of scale, General Schwarzkopf's operations could not compare with, say, those of General Eisenhower in 1944. In point of speed, agility and range, apparently they were no improvement; they were more reminiscent, say, of Eisenhower's 'Broad Front' than of Montgomery's 'Narrow Pencil'. Though detailed information is hard to come by, it has been claimed that, had operations in southern Iraq lasted much longer and penetrated much deeper than the 100 hours they did, they could not have been sustained.[31] Pagonis himself seems to admit as much,[32] albeit only indirectly, as is appropriate for a general in the greatest army on earth that seldom confesses to doing anything less than perfectly.

This brings us to the last question we have to consider, namely the impact that the so-called Revolution in Military Affairs (RMA) can be expected to have on logistics. The way most analysts see it, the RMA got under way around 1990, though some of its origins may be traced to an earlier date. In point of historical significance, it is supposed to be comparable to the introduction of the first effective arquebuses around 1525 and of the first German *Panzer* divisions around 1935.[33] In those times new methods of war enabled those who first adopted them to go from victory to victory, until, that is, everybody else caught up. At the heart of the RMA are very great technological advances in sensors, data links, computers and weapons that can be precision-guided to their targets. A single GPS-guided bomb, or laser-guided air-to-surface missile, can now often substitute for an enormous number of 'dumb' projectiles.

In theory, munitions capable of hitting their targets at the first round should lead to vast logistic savings. Even if no other radical changes take place, the outcome should be faster, deeper-going, more agile operations conducted by much smaller forces. The computers and data links that underpin the RMA can also be used to fine-tune logistic capabilities to operational needs as with civilian just-in-time manufacturing systems, hopefully resulting in additional savings.[34] In practice it is probably too early to tell whether the savings are real. Perhaps the only result is going to be the replacement of mountains of inexpensive old projectiles by mountains of new ones, some of which are very expensive indeed. Is it really cheaper to destroy a target by means of a cruise missile that cost $1 million than by bombing it from the air? The jury on this question is still out, the more so because, as already mentioned, the Second Gulf War was hardly a military contest at all. All it proved was that, when an elephant treads on an ant, the ant is going to be crushed, especially if it does not have a single ally and it has already been crushed before.

In the absence of a comprehensive study, opinions on the effectiveness of just-in-time logistics during the Second Gulf War are divided. Some observers admire the U.S. Army for having covered the 300 miles to Baghdad as fast as it did, and with good reason. On the other hand, *Supplying War* itself shows how Guderian, driving towards Smolensk in 1941, did even better, covering 400 miles in a similar period.[35] Some American soldiers feel they did not obtain sufficient support and criticize the fact that some units went without food for days or found themselves so short of water that they had to buy it from Iraqi vendors they met on the way.[36] Some units had to cannibalize their tanks to keep going; apparently the system that should have permitted every tire and torsion bar to be continuously tracked in real time through every step in the logistic pipeline did not always work as well in practice as in theory.[37]

In relation to the size of the campaign, the consumption of POL was as large as ever, perhaps larger. However, this was one problem that received much command attention; as a result, things worked as well as could be expected and no real shortages developed. Considering the reliance placed on helicopters, which are very maintenance-intensive (particularly in a hot and dusty environment such as the Gulf), the demand for spare parts may also have been greater. Conversely that for ammunition may have been smaller, partly because of improved accuracy but largely because most of Saddam Hussein's troops preferred to throw away their uniforms and disperse rather than face certain destruction. In this respect, too, Operation Iraqi Freedom may have followed a pattern already established during the early years of World War II.

To put it in a different way, while the means and methods used to supply armies have undergone considerable development since 1945, the overwhelming predominance of factory-produced items characteristic of modern war has, if anything, increased. One of the basic arguments of *Supplying War* was precisely that the nature of the supplies required is at least as important as the way of making them available. Hence it does not appear as if the nature of logistics has undergone or is about to undergo a fundamental change; one indication of this is that trucks were as short in 2003 as they had been in 1991 and, indeed, in every single campaign from 1939 on. A real revolution will take place only when soldiers get tired of firing heavy metal projectiles at one another and start using weightless laser beams instead; as to fuel, perhaps the day may come when men, machines and supplies can be beamed around as in the TV series *Star Trek*. In that case it will be vital to have a real-time, three-dimensional, system capable of tracking individual molecules. Its purpose will be to ensure that the colonel, having been dismantled and then reassembled to make his report at the general's headquarters, will not arrive with the sergeant's head screwed to his shoulders by mistake. Now that armies also include women, some of the possibilities are more interesting still.

IV

To sum up, during the thirty years since *Supplying War* was written some things have changed and others have remained much the same. Perhaps the most important single change is the sheer amount of material now available about the history of military logistics and the interest that the subject is capable of raising. Amidst this flood of literature I believe that, as far as it goes and taking the way it defined the subject into account, the book has held up remarkably well. This is true in spite of a few errors it contains, including one on the precise family relationship between Napoleon and Eugene Beauharnais that, to my knowledge, has escaped the notice of every critic so far.

Supplying War argued that the most important turning point in the history of logistics was neither 1789 (gaining freedom from 'umbilical cords or supply' nor 1859–71 (the rise of the railroads) but rather was 1914. Over just a few months, the relationship between food and fodder on the one hand and everything else on the other was inverted. No wonder general staffs everywhere were taken by surprise and that shell shortages, followed by shell scandals, quickly developed. To put it a different way, logistically speaking all warfare took on some of the characteristics of siege warfare. The subsequent development of motor vehicles did restore some of the lost mobility, but the freedom that Alexander or Gustavus Adolphus – even Napoleon, when he decided to march back from Moscow along a route that had not been eaten bare – had enjoyed did not return. If anything, data from the American Civil War, which I did not originally study, reinforce my view. Suppose the statistics are correct and the Union Army really did fire 1,950,000 artillery rounds in the year ending 30 June 1864, as one very reliable source claims.[38] Suppose, too, that the average weight per round had increased 30 percent since the time of Borodino. If so, then a simple calculation will show that, compared to the requirements of about 500,000 men and who knows how many beasts, the weight of ammunition fired still remained trivial.

As to the post-1945 period, there can be little doubt that, in terms of logistics as well as operations, it was not nearly as revolutionary as some people, perhaps seeking to enhance their own accomplishments, have claimed. What the period did see was the continued evolution of a kind of warfare – some analysts call it Third Generation[39] – whose intellectual origins lie as far back as 1918 and which first revealed itself in its full glory in the spring of 1940. As far as logistics go, the most salient single feature of the warfare in question was its dependence on the internal combustion engine both on land and in the air. For decades past, everybody had realized that dependence and some sought to reduce it. As of the present, though, instead of being diminished by new technologies it is still increasing; *vide* the armored divisions that sped to Baghdad in the spring of 2003 and the aircraft and helicopters that protected them – against what? one may well wonder in retrospect – as they did so.

Dependence on the internal combustion engine and, above all, motor vehicles meant that the demand for POL and spare parts continued to increase. The constant upgrading of old weapons and the introduction of some new ones meant that firepower, and with it the demand for ammunition, also increased. Greatly improved aircraft and ships meant that, provided the facilities needed for handling them are available, moving and supplying armed forces worldwide has become much easier and faster; witness, in addition to the American campaigns in the Gulf, the British one in the Falkland Islands. On the other hand, once those forces arrive in theater, and assuming they know anything about how their predecessors did these things, they will be surprised to find how few changes have taken place. Everything else being equal, the growing number of machines that armies drive and fire may have made them more unwieldy and less easy to maneuver than some of their predecessors during World War II. Certainly there is scant evidence of them becoming more mobile and more agile overall; while there are improvements, most of them tend to concern matters of detail.

However that may be, one thing seems certain: in the future, logistics will become even more complex than they already are. In part, this is because they really *are* becoming more complex, what with the introduction of countless new machines and the vast amount of coordination those machines require to keep functioning and fighting. In part, it is because 'complex', in the sense of 'making prolonged (and expensive) study absolutely essential', has become one of the best things one can say about anything. In theory, the use of precision-guided ammunition and the savings to which it might lead should ease some of the logistic burden, the more so because computers can also be used to manage the logistic flow itself. One should keep in mind, though, that theory and practice are nowhere as far apart as in the field of war. To find out why, consult *Supplying War* on the extent to which past armed forces have succeeded in utilizing the maximum theoretical capabilities of the vehicles at their disposal and Clausewitz's *On War* on friction.[40] The latter says that, compared to ordinary life, waging war is like man trying to walk in water: anything that seems easy suddenly requires a much greater effort, and every movement is slowed down.

For a real test of the impact of the RMA on logistics, if any, we shall have to wait until there is a war between the armed forces of two highly developed states; a war that, for excellent nuclear reasons everybody understands, is becoming less likely with every passing day. As Hegel once wrote, Minerva's owl flies at dusk. *Supplying War* was written at a specific point in time, against a specific intellectual background. That background caused it to focus on certain aspects of a certain form of war, one that I myself believe is rapidly becoming a thing of the past. In its place, we witness the proliferation around the world of ballistic missiles, including those that can deliver weapons of mass destruction on the one hand and weapons of terrorism and guerrilla warfare on the other. No doubt these new forms of war will also depend on logistics, and armed organizations still continue to march on their stomachs as they have done since time immemorial. Neither the machines nor the men can function on thin air alone. Even the proverbial handful of rice must be produced, paid for, stored, transported and distributed before it can be consumed; for suicide bombers to blow themselves up, they must be taken to their destination first. The task of exploring the nature of those logistics, as well as of integrating the results with the knowledge that already exists to produce as comprehensive an account as possible, I leave to others. To some of those others, though, I should like to address a warning. Of course there is no question that sound logistics are absolutely essential for the successful conduct of war; in a sense, since precision-guided munitions are even less likely to grow on trees than ordinary ones are, the more modern the war the more true this is. Making sure that you have what you need in the quantity you need it in the place you need it at the time you need it will go a long way towards achieving victory, and the more symmetrical the conflict the more true this is. On the other hand, even in a symmetrical conflict between conventional armed forces, good logistics are not always enough. A *fortiori* in an asymmetrical one: from the British in Palestine to the Russians in Chechnya, much of post-1945 warfare proves that God does *not* always side with the army with more plentiful supplies and the better systems by which to transport and distribute them. Had that been the case, most of the world would still be ruled from London, Paris and a few other colonial capitals. I myself suspect that the Americans in their attempt to police Iraq are going to learn that 'just on time' will not solve all their problems. As well as keeping your powder dry, make sure you keep your faith in God or whatever else you are fighting for. Or else, risk defeat at the hands of lean, mean people who have plenty of nothing and for whom almost nothing is plenty; and who, as sometimes happens, may never have heard the word 'logistics' in the first place.

Note on sources

Owing to the long period and different countries covered by the present book, the sources used and available are extremely varied in nature. Apart from the first chapter, which bears an introductory character, an effort has been made to obtain all the sources available for the study of each campaign; but this was not always possible. In particular, the writing of chapters 3 and 4 (the campaigns of 1870 and 1914) was handicapped by the fact that virtually all the records were destroyed by bombing during World War II. For these chapters, heavy use was therefore made of memoirs and other published material, some of it secondary, for this is all that remains.

While most of the documentary evidence on Napoleon's campaigns of 1805 and 1812 has been published by Alombert-Colin and Fabry, a visit to the *Depôt de la Guerre* at Vincennes brought to light some additional material. For the campaigns of World War II, use has been made of all the archival material I could reach, much of which has never previously been utilized by scholars. That all published material on these campaigns has also been pumped for information is a matter of course.

In the bibliography, apart from a very few exceptions, the only sources listed are those actually quoted in the text. Where English translations of foreign books exist, these have been used. Since each chapter of the book is concerned with only one well-defined period, it was thought best to divide the bibliography accordingly. However, books used for more than one chapter are listed once only.

BIBLIOGRAPHY

Introduction

Glover, R. 'War and Civilian Historians', *Journal of the History of Ideas*, 1957, pp. 91–102

Jomini, A. H. *The Art of War* (Philadelphia, 1873)

Chapter 1

André, L. *Michel Le Tellier et l'Organisation de l'Armée Monarchique* (Paris, 1906)

Idem ed., *Le Testament Politique du Cardinal de Richelieu* (Paris, 1947)

Aubry, Ch. *Le Ravitaillement des Armées de Frédéric Le Grand et de Napoléon* (Paris, 1894)

Audouin, X. *Histoire de l'Administration de la Guerre* (Paris, 1811)

Basta, G. *Il Maestro di Campo Generale* (Venice, 1606)

Bölcker, G. A. *Schola Militaris Moderna* (Frankfurt am Main, 1685)

de Catt, H. *Frederick the Great; the Memoirs of his Reader* (Boston, 1917)

Cherul, M. A., ed., *Lettres de Mazarin* (Paris, 1872)

Clausewitz, C. von *Hinterlassene Werke* (Berlin, 1863)

Delbrück, H. *Geschichte der Kriegskunst im Rahmen der politischen Geschichte* (Berlin, 1920) vol. IV

Dupré d'Aulnay, *Traité générale des subsistances militaires* (Paris, 1744)

Duyck, A. *Journaal* (I. Muller ed., The Hague, 1886)

Fortescue, J. *The Early History of Transport and Supply* (London, 1928)

Greene, W. W. *The Gun and its Development* (London, 1885)

Guibert, J. A. H. de *Essai Générale de Tactique* (Paris, 1803)

Hammerskiöld, 'Ur Svenska Artileriets Hävder', supplement to *Artillerie-Tidskrift*, 1941–44

Hardre, J., ed., *Letters of Louvois* (Chapel Hill, N.C., 1949)

Hondius, H. *Korte Beschrijving ende afbeeldinge van de generale Regelen van de Fortificatie* (The Hague, 1624)

Korn, J. F. *Von den Verpflegungen der Armeen* (Breslau, 1779)

Lefevre, J. *Spinola et la Belgique* (Paris, 1947)

Liddell Hart, B. H. *The Ghost of Napoleon* (London, 1932)

Lorenzen, Th. *Die schwedische Armee im Dreissigjährigen Kriege* (Leipzig, 1894)

Luvaas, J., ed., *Frederick the Great on the Art of War* (New York, 1966)

Marichel, P., ed., *Mémoires du Maréchal du Turenne* (Paris, 1914)

Meixner, O. *Historischer Rückblick auf die Verpflegung der Armeen im Felde* (Vienna, 1895)

Millner, J. *A Compendious Journal of all the Marches, Famous Battles, Sieges...in Holland, Germany and Flanders* (London, 1773)

Monro, R. *His Expeditions with the Worthy Scots Regiment* (London, 1694)

Montecuccoli, R. *Opere* (ed. U. Foscolo, Milan, 1907)

Murray, G., ed., *The Letters and Dispatches of John Churchill, First Duke of Marlborough* (London, 1845)

Parker, G. *The Army of Flanders and the Spanish Road* (Cambridge, 1972)

Perjes, G. 'Army Provisioning, Logistics and Strategy in the Second Half of the 17th Century', *Acta Historica Academiae Scientarium Hungaricae*, No. 16

Pieri, P. 'La Formazione Dottrinale de Raimondo Montecuccoli', *Revue Internationale d'Histoire Militaire*, 10, pp. 92–115

Idem, Principe Eugenio di Savoia, la Campagna del 1706 (Rome, 1936)

Puységur, *Art de la Guerre par principles et règles* (Paris, 1743)

Redlich, F. 'Contributions in the Thirty Years' War', *Economic History Review*, 1959, pp. 247–54

Idem, 'Military Entrepreneurship and the Credit System in the 16th and 17th Centuries', *Kyklos*, 1957, pp. 186–93

Ritter, M. 'Das Kontributionsystem Wallensteins', *Historische Zeitschrift*, 90, pp. 193–247

Roberts, M. *Gustavus Adolphus. A History of Sweden 1611–1632* (London, 1958)

Idem, The Military Revolution (Belfast, 1956)

Rohan, Henri Duc de, *Le Parfacit Capitaine* (Paris, 1636)

Rousset, C. *Histoire de Louvois* (Paris, 1862)

Saxe, M. de *Reveries, or Memoirs on the Art of War* (London, 1957)

Schweren, J. B. J. N. van der, ed., *Brieven en Uitgegeven Stukken van Arend van Dorp* (Utrecht, 1888)

Scoullier, R. E. 'Marlborough's Administration in the Field', *Army Quarterly*, 95, pp. 196–208

Smitt, *Zur näheren Aufklärung über den Krieg von 1812* (Leipzig, 1861)

Spaulding, Q. L., Nickerson, H., and Wright, J. W. *Warfare* (London, 1924)

Stevins, S. *Castramentatio, dat is Legermething* (Rotterdam, 1617)

Taylor, F. *The Wars of Marlborough 1702–1709* (London, 1921)

Tingsten, L. 'Nagra data angaende Gustaf II Adolfs basering och operationsplaner i Tyskland 1630–1632', *Historisk Tidskrift*, 1928, pp. 322–37

Vauban, S. le Pretre de, *Traité des Sièges et del'Attaque des Places* (Paris, 1828)

Vault, F. E., ed., *Mémoires militaires relatifs à la Succession d'Espagne sous Louis XIV* (Paris, 1836)

Wijn, J. W. *Het Krijgswezen in den Tijd van Prins Maurits* (Utrecht, 1934)

Zanthier, F. W. von *Freyer Auszug aus des Herrn Marquis von Santa-Cruz de Marzenado* (Göttingen, 1775)

Chapter 2

I. Unpublished material at the Depôt de la Guerre, Vincennes:

File No.: Title:

C² 3 Correspondance de la Grande Armée, 21–30 September 1805
C² 4 Correspondance de la Grande Armée, 1–10 October 1805
C² 6 Correspondance de la Grande Armée, 21–31 October 1805
C² 8 Correspondance de la Grande Armée, 16–30 November 1805
C² 120 Correspondance de la Grande Armée, 16–31 March 1812
C² 122 Correspondance de la Grande Armée, 1–15 April 1812
C² 524 Correspondance de la Grande Armée, 1812

II. Published

Alombert, P. C. and Colin, J. *La Campagne de 1805 en Allemagne* (Paris, 1902)

Bernardi, *Denkwürdigkeiten aus dem Leben des kaiserlich-russischen Generals von der Infanterie Carl Friedrich von Toll* (Leipzig, 1865)

Brown, N. *Strategic Mobility* (London, 1963)

Bülow, A. von, *Lehrsätze des neueren Krieges oder reine und angewandte Strategie* (Berlin, 1806 ed.)

Cäncrin, *Über die Militärökonomie im Frieden und Krieg* (St Petersburg, 1821)

Caulaincourt, A. de, *With Napoleon in Russia, the Memoirs of Général de Caulaincourt* (New York, 1935)

Chandler, D. *The Campaigns of Napoleon* (New York, 1966)

Chuquet, A. *1812, la Guerre de Russie* (Paris, 1912)

Clausewitz, C. von *The Campaign of 1812 in Russia* (London, 1843)

Colin, J. *L'Education Militaire de Napoléon* (Paris, 1901)

Correspondance de Napoléon Premier (Paris, 1863–)

Dumas, M. *Précis des Evénements Militaires 1799–1814* (Paris, 1822)

Fabry, L. G. *Campagne de Russie 1812* (Paris, 1912)

Fortescue, J., ed., *Notebook of Captain Coignet* (London, 1928)

Grouard, ed., *Mémoirs écrits en Sainte Hélène* (Paris, 1822–)

Hyatt, A. H. J. 'The Origins of Napoleonic Warfare; a Survey of Interpretations', *Military Affairs*, 1966, pp. 177–85

Krauss, A. *Der Feldzug von Ulm* (Vienna, 1912)

Montholon, C. J. F. T., ed., *Recits de la captivité de l'empereur Napoléon à Sainte Hélène* (Paris, 1847)

Nanteuil, H. de, *Daru et l'Administration militaire sous la Révolution et l'Empire* (Paris, 1966)

Perjes, G. 'Die Frage der Verpflegung im Feldzuge Napoleons gegen Russland', *Revue Internationale d'Histoire Militaire*, 1968, pp. 203–31

Picard ed., *Napoléon, Précepts et Jugements* (Paris, 1913)

Quimby, A. *The Background to Napoleonic Warfare* (New York, 1957)

Ségur, P. de, *Histoire de Napoléon et de la Grande Armée pendant l'année 1812* (n.p, n.d)

Tarlé, E. *Napoleon's Invasion of Russia 1812* (London, 1942)

Tulard, J. 'La Depôt de la Guerre et la préparation de la Campagne en Russie', *Revue Historique de l'Armée*, 1969

Ullmann, J. *Studie über die Ausrüstung sowie über das Verpflegs- und Nachschubwesen im Feldzug Napoleon I gegen Russland im Jahre 1812* (Vienna, 1891)

Vie Politique et Militaire de Napoléon, racontée par lui même (Brussels, 1844)

Chapter 3

Anon, *De l'Emploi des Chemins de Fer en Temps de Guerre* (Paris, 1869)

Blumenthal, *Tagebücher* (Stuttgart, 1902)

Boehn, H. von *Generalstabgeschäfte; ein Handbuch für Offiziere aller Waffen* (Potsdam, 1862)

Bondick, E. *Geschichte des Ostpreussischen Train-Battalion Nr. 1* (Berlin, 1903)

Budde, H. *Die französischen Eisenbahnen in deutschen Kriegsbetriebe 1870/71* (Berlin, 1904)

Clausewitz, C. von *On War* (London, 1904)

Craig, G. A. *The Battle of Königgratz* (London, 1965)

Engelhardt W. 'Rückblicke auf die Verpflegungsverhältnisse im Kriege 1870/71', Beiheft 11 zum *Militärwochenblatt*, 1901

Ernouf, *Histoire des Chemins de Fer Français pendant la Guerre Franco-Prussienne* (Paris, 1874)

François, H. von *Feldverpflegung bei der höheren Kommandobehörden* (Berlin, 1913)

Fransecky, E. von *Denkwürdigkeiten* (Berlin, 1913)

Goltz, C. von der 'Eine Etappenerinnerung aus dem Deutsch-Französischen Kriege von 1870/71', Beiheft zum *Militärwochenblatt*, 1866

Heine, W. 'Die Bedeutung der Verkehrswege für Plannung und Ablauf militärischer Operationen', *Wehrkunde*, 1965, pp. 424–9

von Hesse, 'Der Einfluss der heutigen Verkehrs- und Nachrichtenmittel auf die Kriegsführung', Beiheft zum *Militärwochenblatt*, 1910

Hille & Meurin, *Geschichte der preussischen Eisenbahntruppen* (Berlin, 1910)

Hold, A. 'Requisition und Magazinsverpflegung während der Operationen', *Organ der Militär-Wissenschaftlichen Vereine*, 1878, pp. 405–87

Howard, M. *The Franco-Prussian War* (London, 1961)

Hozier, H. M. *The Seven Weeks' War* (London and New York, 1871)

Jacqmin, F. *Les Chemins de fer pendant la guerre de 1870–71* (Paris 1872)

Kaehne, H. *Geschichte des Königlich Preussischen Garde-Train-Battalion* (Berlin, 1903)

Layritz, O. *Mechanical Traction in War* (Newton Abbot, Devon, 1973 reprint)

Lehmann, G. *Die Mobilmachung von 1870/71* (Berlin, 1905)

List, F. 'Deutschlands Eisenbahnsystem im militärischer Beziehung', in *Schriften, Reden, Briefe* (E. von Beckerath and O. Stühler eds., Berlin, 1929)

Luard C. E., 'Field Railways and their General Application in War', *Journal of the Royal United Services Institute*, 1873, pp. 693–715

McElwee, W. *The Art of War, Waterloo to Mons* (London, 1974)

Meinke, B. 'Beiträge zur frühesten Geschichte des Militäreisenbahnwesens', *Archiv für Eisenbahnwesen*, 1938, pp. 293–319

Molnar, H. von 'Uber Ammunitions-Ausrüstung der Feld Artillerie', *Organ der Militär-Wissenschaftlichen Vereine*, 1879, i, pp. 585–614

Moltke, H. von, *Dienstschriften* (Berlin, 1898)
Idem, Militärische Werke (Berlin, 1911)
Niox, M. *De l'Emploi des Chemins de Fer pour les Mouvements
Stratégiques* (Paris, 1873)
Pönitz, K. E. von, *Die Eisenbahnen und ihre Bedeutung als militärische
Operationslinien* (Adorf, 1853)
Roginat, A. de, *Considérations sur l'art de la Guerre* (Paris, 1816)
Schreiber, *Geschichte des Brandenburgishen Train-Battalions Nr. 3* (Berlin,
1903)
Shaw, G. C. *Supply in Modern War* (London, 1938)
Showalter, D. E. 'Railways and Rifles; the Influence of Technological
Developments on German Military Thought and Practice, 1815–1865'
(University of Minnesota Dissertation, 1969)
Stoffel, *Rapports Militaires 1866–70* (Paris, 1872)
Whitton, F. E. *Moltke* (London, 1921)
Wilhelm, Rex, *Militärische Schrifte* (Berlin, 1897)
H. L. W., *Die Kriegführung unter Benützung der Eisenbahnen* (Leipzig,
1868)

Chapter 4

Addington, L. H. *The Blitzkrieg Era and the German General Staff,
1865–1941* (New Brunswick, N.J., 1971)
Anon, *Der Krieg, Statistisches, Technisches, Wirtschaftliches* (Munich, 1914)
Anon, 'Die kritische Transportweite im Kriege', *Zeitschrift für
Verkehrwissenschaft*, 1955, pp. 119–24
Army/USA/War College, 'Analysis of the Organization and Administration
of the Theater of Operations of the German 1. Army in 1914', unpublished
analytical study (Washington, D.C., 1931)
Army/USA/War College, 'Analytical Study of the March of the German 1.
Army, August 12–24 1914', unpublished staff study (Washington, D.C.,
1931)
Asprey, R. *The Advance to the Marne* (London, 1962)
Baumgarten-Crusius, A. von, *Deutsche Heeresführung im Marnefeldzug
1914* (Berlin, 1921)
Behr, H. von, *Bei der fünften Reserve Division im Weltkrieg* (Berlin, 1919)
Bernhardi, F. von, *On War of Today* (London, 1912)
Binz, G. L., 'Die stärkere Battalione', *Wehrwissenschaftliche Rundschau*,
9, pp. 63–85
Bloem, W. *The Advance from Mons* (London, 1930)
Cochenhausen, F. von, *Heerführer des Weltkrieges* (Berlin, 1939)
Earle, E. M. *Makers of Modern Strategy* (New York, 1970 reprint)
von Falkenhausen, *Der grosse Krieg der Jetztzeit* (Berlin, 1909)
Flammer, Ph. M. 'The Schlieffen Plan and Plan XVII; a Short Critique',
Military Affairs, 30, pp. 207–12
Förster, W. *Graff Schlieffen und der Weltkrieg* (Berlin, 1921)
Föst, *Die Dienst der Trains im Kriege* (Berlin, 1908)
François, H. von, *Marneschlacht und Tannenberg* (Berlin, 1920)
Gackenholtz, H. *Entscheidung in Lothringen 1914* (Berlin, 1933)

Goltz, C. von der, *The Nation in Arms* (London, 1913)

Gröner, W. *Der Feldherr wider Willen* (Berlin, 1931)

Idem, Lebenserinnerungen (Göttingen, 1957)

Hauesseler, H. *General Wilhelm Gröner and the Imperial German Army* (Madison, Wisc., 1962)

Haussen, M. K. von. *Erinnerungen an der Marnefeldzug 1914* (Leipzig, 1920)

Helfferich, K. von, *Der Weltkrieg* (Berlin, 1919)

Henniker, A. M. *Transportation on the Western Front, 1914–1918* (London, 1937)

Heubes, M. *Ehrenbuch der deutschen Eisenbahner* (Berlin, 1930)

Jäscke, G. 'Zum Problem der Marne-Schlacht von 1914', *Historische Zeitschrift*, 190, pp. 311–48

Jochim, Th. *Die Operationen und Rückwärtigen Verbindungen der deutsche 1. Armee in der Marneschlacht 1914* (Berlin, 1935)

Justrow, K. *Feldherr und Kriegstechnik* (Oldenburg, 1933)

Kabisch, E. *Streitfragen des Weltkrieges 1914–1918* (Stuttgart, 1924)

Kluck, A. von, *Der Marsch auf Paris und die Marneschlacht 1914* (Berlin, 1920)

Kretschmann, W. *Die Wiederherstellung der Eisenbahnen auf dem westlichen Kriegsschauplatz* (Berlin, 1922)

Kuhl, H. von, *Der deutsche Generalstab in Vorbereitung und Durchführung des Weltkrieges* (Berlin, 1920)

Idem, and Bergmann, J. von, *Movements and Supply of the German First Army during August and September 1914* (Fort Leavenworth, Kan., 1920)

Ludendorff, E. *The General Staff and its Problems* (Berlin, 1920)

Marx, W. *Die Marne, Deutschlands Schicksal?* (Berlin, 1932)

Moltke, H. von, *Die deutsche Tragödie an der Marne* (Berlin, 1922)

Idem, Erinnerungen, Briefe, Dokumente 1877–1916 (Berlin, 1922)

Müller-Löbnitz, W. *Die Sendung des Oberstleutnants Hentsch am 8–10 September 1914* (Berlin, 1922)

Napier, C. S. 'Strategic Movement by Rail in 1914', *Journal of the Royal United Services Institute*, 80, pp. 69–93, 361–80

Idem, 'Strategic Movement over Damaged Railways in 1914', *ibid*, 81, pp. 315–46

OHL ed., *Militärgeographische Beschreibung von Nordost Frankreich, Luxemburg, Belgien und dem südlichen Teil der Niederlande und der Nordwestlichen Teil der Schweiz* (Berlin, 1908)

Idem, Taschenbuch für den Offizier der Verkehrstruppen (Berlin, 1913)

Poseck, M. von *Die deutsche Kavallerie 1914 in Belgien und Frankreich* (Berlin, 2nd ed., 1921)

Pratt, E. A. *The Rise of Rail-Power in War and Conquest, 1833–1914* (London, 1916)

Reichsarchiv ed., *Der Weltkrieg* (Berlin, 1921–)

Idem, Der Weltkrieg, das deutsche Feldeisenbahnwesen (Berlin, 1928)

Riebecke, O. *Was brauchte der Weltkrieg?* (Berlin, 1936)

Ritter, G. *The Schlieffen Plan, Critique of a Myth* (London, 1957)

Rochs, H. *Schlieffen, Ein Lebens– und Charakterbild für das deutsche Volk* (Berlin, 2nd ed., 1921)

Schlieffen, A. von, *Gesammelte Schriften* (Berlin, 1913)

Schwarte, M. *Die militärische Lehre des grossen Krieges* (Berlin, 1923)
Tappen, A. von, *Bis zur Marne* (Oldenburg, 1920)
Villate, R. 'L'Etat matériel des armées allemandes en Août et Septembre 1914', *Revue d'Histoire de la Guerre Mondiale*, 4, pp. 310–26
Voorst tot Voorst, J. J. G. van, 'Over Roermond!' appended to *De Militaire Spectateur*, 92
Wallach, J. L. *Das Dogma der Vernichtungsschlacht* (Frankfurt am Main, 1967)
Wrissberg, E. *Heer und Waffen* (Leipzig, 1922)

Chapter 5

I. Unprinted[1]
A. At the Militärgeschichtliches Forschungsamt, Freiburg i.B.:
    AOK 2/0.Qu, Anlagen zum KTB, files 16773/14
    AOK 4/0.Qu, Anlagen zum KTB, files 17847/3, 17847/4, 17847/5
    AOK 9/0.Qu, KTB, file 13904/1, Anlagen zum KTB, files 13904/2, 13904/4, 13904/5
    Aussenstelle OKH/Gen.Qu/HGr. Sud, KTB, file 27927/1, Anlagen zum KTB, file 27927/8
    Pz.Gr. 1/0.Qu, KTB, file 16910/46
    Pz.Gr. 2/0.Qu, KTB, file RH 21–2/v819, Anlagen zum KTB, files RH 21–2/v823, RH 21–2/v829, H 10–51/2
    Pz.Gr. 4/0.Qu, KTB, file 22392/1, Anlagen zum KTB, files 13094/5, 22392/22, 22392/23.
B. On microfilm:[2]
    Aussenstelle OKH/Gen.Qu/HGr. Nord, KTB, GMR/T–311/111/714934
    Aussenstelle OKH/Gen.Qu./HGr. Sud, Anlagen zum KTB, GMR/T–311/264/000365
    Eisenbahnpionierschule, Folder with various documents, GMR/T–78/259/6205611
    Grukodeis C, Folder with various documents, GMR/T–78/259/6204883
    Unknown, Various documents re the *Eisenbahntruppe*, GMR/T–78/259/6204125
II. Published
Beaulieu, W. C. de *Generaloberst Erich Hoepner* (Neckargemund, 1969)
Idem, 'Sturm bis Moskaus Tore', *Wehrwissenschaftliche Rundschau*, 1956, pp. 349–65, 423–39

---

[1] In the original records there is often some ambiguity over the exact title of this or that headquarters. Thus, Aussenstelle OKH/Gen.Qu/HGr. Sud was sometimes known as Gen.Qu/Aussenstelle Sud. Pz.Gr. 2 was sometimes known as Armeegruppe Guderian. To avoid confusion, these differences have been disregarded and a uniform nomenclature employed throughout.
[2] Some of the folders do not have proper titles, and those given (in English) are just a rough indication of contents.

Bork, M. 'Das deutsche Wehrmachttransportwesen – eine Vorstufe europäischer VerkehrsFührung', *ibid*, 1952, pp, 50–6

Cecil, R. *Hitler's Decision to Invade Russia 1941* (London, 1975)

Creveld, M. van *Hitler's Strategy 1940–41; the Balkan Clue* (Cambridge, 1973)

Idem, 'Warlord Hitler; Some Points Reconsidered', *European Studies Review*, 1974, pp. 557–79

Friedenburg, F. 'Kan der Treibstoffbedarf der heutigen Kriegsführung überhaupt befriedigt werden?' *Der deutsche Volkswirt*, 16 April 1937

Guderian, H. *Panzer Leader* (London, 1953)

Halder, F. *Kriegstagebuch* (Stuttgart, 1961–3)

Haupt, W. *Heeresgruppe Nord 1941–45* (Bad Nauheim, 1966)

Hillgruber, A. *Hitlers Strategie* (Frankfurt am Main, 1965)

International Military Tribunal, *Trial of the Major War Criminals* (Nuremburg, 1946–)

Jacobsen, H. A. *Fall Gelb* (Wiesbaden, 1957)

Idem, 'Motorisierungsprobleme im Winter 1939/40', *Wehrwissenschaftliche Rundschau*, 1956, pp. 497–518

Kreidler, E. *Die Eisenbahnen im Machtbereich der Assenmächte wahrend des Zweiten Weltkrieges* (Göttingen, 1975)

Kriegstagebuch des Oberkommando der Wehrmacht (Frankfurt am Main, 1965)

Krumpelt, I. *Das Material und die Kriegführung* (Frankfurt am Main, 1968)

Idem, 'Die Bedeutung des Transportwesens für den Schlachterfolg', *Wehrkunde*, 1965, pp. 465–72

Leach, B. *erman Strategy against Russia* (London, 1973)

Liddell Hart, B. H. *History of the Second World War* (London, 1973)

Mueller-Hillebrand, B. *Das Heer* (Frankfurt am Main, 1956)

Overy, R. J. 'Transportation and Rearmament in the Third Reich' *The Historical Journal*, 1973, pp. 389–409

Ploetz, A. G. *Geschichte des Zweiten Weltkrieg* (Würzburg, 1960)

Pottgiesser, H. *Die deutsche Reichsbahn im Ostfeldzug 1939–1944* (Neckargemund, 1960)

Rohde, H. *Das deutsche Wehrmachttransportwesen im Zweiten Weltkrieg* (Stuttgart, 1971)

Seaton, A. *The Russo-German War* (London, 1971)

Steiger, R. *Panzertaktik* (Freiburg, i.B. 1973)

Teske, H. *Die silbernen Spiegel* (Heidelberg, 1952)

Thomas, G. *Geschichte der deutschen Wehr- und Rüstungswirtschaft 1918–1943/45* (Boppard am Rhein, 1966)

Trevor-Roper, H. R. *Hitler's War Directives* (London, 1964)

Wagner, E. *Der Generalquartiermeister* (Munich and Vienna, 1963)

Windisch, *Die deutsche Nachschubtruppe im Zweiten Weltkrieg* (Munich, 1953)

Chapter 6

I. Unpublished, on microfilm
    DAK/Qu, KTB with Anlagen, GMR/T–314/15/000861

DAK/Qu, KTB with Anlagen, GMR/T–314/16

OKH/Genst.d.H/Eremde Heere West, Intelligence reports, North Africa, GMR/T–78/45/6427915

OKH/Genst.d.H/Op.Abt. (IIb), Chefsachen-Feindlbeurteilungen und eigene Absichten, GMR/T–78/646/0000933

OKH/Genst.d.H/Op.Abt. (III & IIb), Meldungen Nordafrika, GMR/T–78/325

OKH/Genst.d.H/Op.Abt. (IIb), Tagesmeldungen Nord-Afrika, GMR/T–78/324

OKH/Genst.d.H/Org.Abt. (III), KTB, GMR/T–78/414/6382356

Stato Maggiore Generale, Verbale del Colloquio tra l'Eccelenza Cavallero ed il Maresciallo Keitel, 25.8.1941, IMR/T–821/9/000322

Also on microfilm:

B. Mueller-Hillebrand, 'Germany and her Allies in World War II' unpublished study, US Army Historical Division MS No. P–108, GMR/63–227

II. Published

Assmann, K. *Deutsche Schicksahljahre* (Wiesbaden, 1951)

Baum, W. and Weichold, E. *Der Krieg der Assenmächte im Mittelmeer-Raum* (Göttingen, 1973)

Becker, C. *Hitler's Naval War* (London, 1974)

Bernotti, R. *Storia della guerra nel Mediterraneo* (Rome, 1960)

Bragadin, M. A. *The Italian Navy in World War II* (Annapolis, Md., 1957)

Brant, E. D. *Railways of North Africa* (Newton Abbot, Devon, 1971)

Cavallero, U. *Commando Supremo* (Bologna, 1948)

Detwiller, D. S. *Hitler, Franco und Gibraltar* (Wiesbaden, 1962)

Documents on German Foreign Policy (Washington, D.C., and London, 1948–), series D. vol. xi

Faldella, E. *L'Italia e la seconda guerra mondiale* (Bologna, 1959)

Favagrossa, C. *Perchè perderemo la guerra* (Milan, 1947)

Gabriele, M. 'La Guerre des Convois entre l'Italie et l'Afrique du Nord', in Comité d'Histoire de la Deuxième Guerre Mondiale ed., *La Guerre en Mediterranée 1939–1945* (Paris, 1971)

Gruchman, L. 'Die "verpassten strategischen chancen" der Assenmächte, im Mittelmeeraum 1940–1941', *Vierteljarshefte für Zeitgeschichae*, 1970, pp. 456–75

Jacobsen, H. A. and Rohwehr, J. *Decisive Battles of World War II* (London, 1965)

Jäckel, E. *Frankreich in Hitlers Europa* (Stuttgart, 1966)

Kesselring, A. *Memoirs* (London, 1953)

Les Lettres secrètes échangées par Hitler et Mussolini (Paris, 1946)

Liddell Hart, B. H., *The German Generals Talk* (London, 1964)

Idem, ed., *The Rommel Papers* (New York, 1953)

Macintyre, D. *The Battle for the Mediterranean* (London, 1964)

Maravigna, R. *Come abbiamo perduto la guerra in Africa* (Rome, 1949)

Playfair, S. O. *The Mediterranean and the Middle East* (London, 1956)

Rintelen, E. von 'Operation und Nachschub', *Wehrwissenschaftliche Rundschau*, 1951, 9–10, pp. 46–51

Rochat, G. 'Mussolini Chef de Guerre', *Revue d'histoire de la deuxième guerre mondiale*, 1975, pp. 59–79

Rosskill, S. W. *The War at Sea 1939–1945* (London, 1956)

Stark, W. 'The German Afrika Corps', *Military Review, 1965*, pp. 91–7

Stato Maggiore Esercito/Ufficio Storico ed., *Terza Offensiva Britannica in Africa Settentrionale* (Rome, 1961)

Westphal, S. *Erinnerungen* (Mainz, 1975)

Idem ed., *The Fatal Decisions* (London, 1956)

Weygand, M. *Recalled to Service* (London, 1952)

Chapter 7

I. Unprinted

A. Material at States House, Medmenham, Bucks (the Liddell Hart Centre for Military Archives): file, No. 15/15/30, 15/15/47, 15/15/48 of the Chester Wilmot Collection

B. Files at the Public Record Office, London:
WO/171/146, WO/171/148, WO/205/247, WO/219/259, WO/219/260, WO/219/2521, WO/219/2976, WO/219/3233.

II. Printed

A. American Official Histories

Bykofsky, J. and Larson, H. *The Transportation Corps* (*Washington, D.C.*, 1957)

Harrison, G. A. *Cross Channel Attack* (Washington, D.C., 1951)

Ross, W. F. and Romanus, C. F. *The Quartermaster Corps; Operations in the War Against Germany* (Washington, D.C., 1965)

Ruppenthal, R. G. *Logistical Support of the Armies* (Washington, D.C., 1953)

B. British Official Histories

Ellis, H. *Victory in the West* (London, 1953)

Ehrman, J. *Grand Strategy* (London, 1956) vol. v

C. Other printed sources

Administrative History, 21. Army Group (London, n.d.)

Bradley, O. N. *A Soldier's Story* (London, 1951)

Busch, E. 'Quartermaster Supply of Third Army', *The Quartermaster Review*, November–December 1946, pp. 8–11, 71–8

Butcher, H. C. *Three Years with Eisenhower* (London, 1946)

Chandler, A. D., ed., *The Papers of Dwight David Eisenhower* Baltimore, Md., 1970)

Eisenhower, D. D. *Crusade in Europe* (London, 1948)

Greenfield, K. R., ed., *Command Decisions* (London, 1960)

Huston, J. A. *The Sinews of War* (Washington, D.C., 1966)

Liddell Hart, B. H. 'Was Normandy a Certainty?' in *Defense of the West* (London, 1950)

Montgomery, B. L. *Memoirs* (London, 1958)

Morgan, F. *Overture to Overlord* (London, 1950)

'Musketeer', 'The Campaign in N.W. Europe, June 1944–February 1945', *Journal of the Royal United Services Institute*, 1958, pp. 72–81

Patton, G. S. *War as I Knew It* (London, n.d.)
Wilmot, C. *The Struggle for Europe* (London, 1952)

Chapter 8

Leighton, R. M. 'Logistics' in *Encyclopaedia Britannica,* 14th edition (London and New York, 1973)
Muraise, E. *Introduction à l'Histoire militaire* (Paris, 1964)
Wavell, A. C. P. *Speaking Generally* (London, 1946)
Wheldon, J. *Machine Age Armies* (London, 1968)

NOTES

Introduction

1 A. H. Jomini, *The Art of War* (Philadelphia, 1873) p. 225.
2 R. Glover, 'War and Civilian Historians', *Journal of the History of Ideas*, 1957, p. 91.
3 See his introduction to G. Riter, *The Schlieffen Plan* (London, 1957) pp. 6–7.

Chapter 1

1 For detailed figures see G. Parker, *The Army of Flanders and the Spanish Road* (Cambridge, 1972) p. 28.
2 S. Stevins, *Castramentatio, dat is Legermething* (Rotterdam, 1617) pp. 25–7.
3 See J. W. Wijn, *Het Krijgswezen in den Tijd van Prins Maurits* (Utrecht, 1934) p. 386. The Spaniards were even worse off; in 1606, Spinola had 2,000–2,500 wagons for 15,000 men.
4 For a good example of the duties involved in the job see the Instruction appointing Arend van Dorp as Superintendent to the Army of François Alençon in 1582, printed in J. B. J. N. van der Schweren ed., *Brieven en Uitgegeven Stukken van Arend van Dorp* (Utrecht, 1888) ii, 2–9.
5 G. Basta, *Il Maestro di Campo Generale* (Venice, 1600) p. 8.
6 Decrees limiting the number of sutlers per company were often issued but seldom enforced.
7 The prime example of such a route was the famous 'Spanish road', used by generations of commanders to march troops from Italy to the Netherlands. For its organization see Parker, *op. cit.*, pp. 80–105.
8 This was an important point: see e.g. the Duc de Rohan, *Le Parfaict Capitaine* (Paris, 1636) p. 198.
9 See F. Redlich, 'Military Entrepreneurship and the Credit System in the 16th and 17th Centuries', *Kyklos*, 1957, 186–93.
10 For Sully see X. Audouin, *Histoire de l'Administration de la Guerre* (Paris, 1811) ii. p. 39ff; for Spinola J. Lefevre, *Spinola et la Belgique* (Paris, 1947) pp. 45–51.

11 On Wallenstein's supplies see M. Ritter, 'Das Kontributionsystems Wallensteins', *Historische Zeitschrift*, 90, pp. 210–11; also F. Redlich, 'Contributions in the Thirty Years' War', *Economic History Review*, 1959, pp. 247–54.

12 Wijn, *op. cit.*, pp. 383–5.

13 E.g. Spinola's conquest of Friesland in 1605.

14 See M. Roberts, *The Military Revolution* (Belfast, 1956) pp. 15–16.

15 H. Hondius, *Korte Beschrijving ende afbeeldinge van de generale Regelen van de Fortificatie* (The Hague, 1624) p. 43. A *last* equalled 4480 lb.

16 Maurice's campaign is described in detail by A. Duyck, *Journaal* (I. Muller ed., The Hague, 1886) iii, 384, 389ff.

17 For detailed lists down to the last spike see Stevins, *op. cit.*, p. 45ff.

18 M. Roberts, *Gustavus Adolphus, A History of Sweden 1611–1632* (London, 1958) ii, 228–34, 270.

19 In January 1629 Wallenstein wrote that 'on the island of Rügen the troops readily eat dogs and cats, whereas the peasants simply drown themselves in the sea out of hunger and desperation'.

20 Roberts, *Gustavus Adolphus*, ii, 471.

21 Quoted in Th. Lorenzen, *Die schwedische Armee im Dreissigjährigen Kriege* (Leipzig, 1894) pp. 22–3.

22 L. Hammerskiöld, 'Ur Svenska Artilleriets Hävder', supplement to *Artillerie-Tidskrift*, 1941–4, p. 169.

23 See the description in R. Monro, *His Expeditions with the Worthy Scots Regiment* (London, 1634) ii, 89.

24 Roberts, *Gustavus Adolphus*, ii, pp. 676–7.

25 See L. Tingsten, 'Nagra data angaende Gustaf II Adolfs basering och operationsplaner i Tyskland 1630–32', *Historisk Tidskrift*, 1928, pp. 322–37.

26 L. André ed., *Le Testament Politique du Cardinal de Richelieu* (Paris, 1947) p. 480.

27 On the general character of warfare in this period see P. Pieri, 'La Formazione Dottrinale de Raimondo Montecuccoli', *Revue Internationale d'Histoire Militaire*, 1951, esp. pp. 98–100.

28 Rohan, *op. cit.*, pp. 331–2.

29 See M. A. Cherul ed., *Lettres de Mazarin* (Paris, 1872) vols. ii, iii, *passim*. For Le Tellier's reforms in general, L. André, *Michel Le Tellier et l'Organisation de l'Armée Monarchique* (Paris, 1906) p. 457ff.

30 P. Marichel ed., *Mémoires du Maréchal du Turenne* (Paris, 1914) ii, pp. 115, 153ff.

31 Audouin, *op. cit.*, ii, 236–7.

32 For an example of the way in which the soldiers' consumption was calculated in this period see R. Montecuccoli, *Opere* (ed. U. Foscolo, Milan, 1807) pp. 46–7.

33 C. Rousset, *Histoire de Louvois* (Paris, 1862) i, 1, p. 249ff.

34 Louvois to Montal, 18 August 1683, printed in J. Hardre ed., *Letters of Louvois* (Chapel Hill, N.C., 1949) pp. 234–5.

35 Louvois to Plesis, 8 June 1684; Louvois to Charuel, 11 June 1684, *ibid*, pp. 362, 364.

36 Louvois to Humiers, 21 September 1683, *ibid*, p. 265.

37 Louvois to Bellefonds, 21 March 1684, *ibid*, pp. 489–90.
38 Puységur, *Art de la guerre par principles et règles* (Paris, 1743) ii, p. 64, gives the ratio as 3:2.
39 J. A. H. de Guibert, *Essai Général de Tactique* (Paris, 1803) II, pp. 265–6; B. H. Liddell Hart, *The Ghost of Napoleon* (London, 1932) p. 27ff.
40 Louvois to de la Trousse, 17 May 1684; Louvois to Breteuil, 25 April 1684; Hardre, *op. cit.*, 352–3, 347.
41 De Boufflers to Louis XIV, 22 June 1702, printed in F. E. Vault ed., *Mémoires militaires relatifs à la Succession d'Espagne sous Louis XIV* (Paris, 1836) i, 59–61; Houssaye memorandum, 17 July 1705, *ibid*, v, 790–4.
42 Puységur to Chamillart, April 1702, *ibid*, ii, 17–19.
43 Louvois to Chamlay, 12 June 1684, Hardre, *op. cit.*, pp. 366–7.
44 G. Perjes, 'Army Provisioning, Logistics and Strategy in the Second Half of the 17th Century', *Acta Historica Academiae Scientarium Hungaricae*, 16, p. 27 and footnote 64.
45 'The lines were seized, and the communication with Holland interrupted. The Duke opened new communications with great labour, and greater skill, through countries overrun by the enemies. The necessary convoys arrived in safety.'
46 E.g. the sieges of Bruges and Mons; see Louvois' letters in Hardre, *op. cit.*, pp. 314–15, 337.
47 E.g. Liddell Hart, *op. cit.*, pp. 23–7; O. L. Spaulding, H. Nickerson and J. W. Wright, *Warfare* (London, 1924) p. 550ff.
48 Guibert, *op. cit.*, ii, 295.
49 See H. Delbrück, *Geschichte der Kriegskunst in Rahmen der politische Geschichte* (Berlin, 1920) iv, *passim*.
50 M. de Saxe, *Reveries, or Memoirs upon the Art of War* (London, 1957) p. 8.
51 F. W. von Zanthier, *Freyer Auszug aus des Herrn Marquis von Santa-Cruz de Marzenado* (Göttingen, 1775) pp. 56–7.
52 Attributed to Frederick II and quoted in H. de Catt, *Frederick the Great; the Memoirs of his Reader* (Boston, 1917) ii, 223.
53 J. F. Korn, *Von den Verpflegungen der Armeen* (Breslau, 1779) p. 20.
54 K. von Clausewitz, *Hinterlassene Werke* (Berlin, 1863) x, p. 69.
55 Korn, *op. cit.*, pp. 82–6. J. Luvaas ed., *Frederick the Great on the Art of War* (New York, 1966) pp. 110–11; Ch. Aubry, *Le Ravitaillement des Armées de Frédéric le Grand et de Napoléon* (Paris, 1894) p. 13.
56 Clausewitz, *On War*, ii, p. 87ff.
57 G. Murray ed., *The Letters and Dispatches of John Churchill, First Duke of Marlborough* (London, 1845) i, 291–92.
58 J. Fortescue, *The Early History of Transport and Supply* (London, 1928) p. 12.
59 Murray, *op. cit.*, i, 226, 240.
60 *Ibid*, 289, 311.
61 E.g. Marlborough to the Magistrates of Schrobenhausen, 24 July 1704, *ibid*, 371–2.

62 Marlborough to Harley, 28 August 1704, *ibid*, 437; J. Millner, *A Compendious Journal of all the Marches, Famous Battles, Sieges. . .in Holland, Germany and Flanders* (London, 1773) p. 132; also R. E. Scoullier, 'Marlborough's Administration in the Field', *Army Quarterly*, 95, p. 203.

63 Murray, *op. cit.*, i, 301.

64 *Ibid*, 363, 367–8.

65 *Ibid*, 368, 370.

66 *Ibid*, 382; also F. Taylor, *The Wars of Marlborough 1702–1709* (London, 1921) i, 187–96.

67 Millner, *op. cit.*, p. 111.

68 E.g. during the siege of Landau, Murray, *op. cit.*, i, 497.

69 Dupré d'Aulnay, *Traité générale des subsistances militaires* (Paris, 1744) pp. 150–60.

70 See e.g. the description in Luvaas, *op. cit.*, pp. 111–13; also Perjes, *loc. cit.*, pp. 17–19.

71 O. Meixner, *Historischer Rückblick auf die Verpflegung der Armeen im Felde* (Vienna, 1895) i, part i, 23–4.

72 Based on Perjes, *loc. cit.*, pp. 4–5.

73 These figures are based on Puységur and are correct for the beginning of the century; in the 'seventeen forties, Dupré d'Aulnay was reckoning only 80,000 rations for an army with 60,000 men.

74 For western Europe at any rate, this is much too low; here density of population ranged between about thirty-five (Prussia) and 110 (Lombardy) per square mile.

75 Puységur, *op. cit.*, ii, 64.

76 640 acres equal one square mile.

77 On the assumption that the inhabitants used the foraging areas twice as effectively as the troops.

78 I have not been able to find figures on the number of horses as against that of people in this period; in 1812, however, general Pfuel calculated it as approximately 1:3 for the poor regions of western Russia. See Smitt, *Zur näheren Aufklärung über den Krieg von 1812* (Leipzig, 1861) p. 439ff.

79 Perjes, *loc. cit.*, p. 26ff.

80 See below, pp. 102–3.

81 E.g. the French sieges of Bruges and Mons in 1683–84; Hardre, *op. cit.*, pp. 314–15, 337.

82 Rohan, *op. cit.*, p. 318; Montecuccoli, *op. cit.*, pp. 45–6; G. A. Bölcker, *Schola Militaris Moderna* (Frankfurt am Main, 1685) p. 68.

83 To capture a 'strong' fortress in one month, Vauban (*Traité des Siéges et del'Attaque des Places*, Paris, 1828, i, 14–15) calculates 60,000 infantry and 16,000 artillery rounds, the latter to be fired by 130 guns of various calibres. Assuming 12 infantry rounds to the pound and an average of 12lb per cannon-ball, overall weight of the ammunition must have come (including powder at fifty per cent of rounds fired) to 132 tons.

Vauban does not give the strength of the army he has in mind; in 1692, however, Louis XIV besieged Mons with 60,000 men and 151 cannon, while another 60,000 troops were stationed some distance away to prevent relief. Assuming a very modest 40,000 men for Vauban's 130 guns, consumption of provisions during a month's siege was 1,600 tons, more than ten times the weight of the ammunition. If the requirements of the 20,000 horses that were likely to accompany such a force are added, the share of ammunition dwindles to less than two per cent.

84 W. Greene, *The Gun and its Development* (London, 1885) p. 45.

85 Clausewitz, *On War*, books i, iv; Delbrück, *op. cit.*, vol. iv *passim*.

86 The marches of Prince Eugene, Marlborough and Frederick II have already been mentioned. 'Free from the magazine system' as they were, the former two were outmarched on their way to Toulon by a French general, Tesse; while other French commanders were able to march from Strasbourg to Bavaria at a pace not at all inferior to that of the great Englishman.

87 See Scoullier, *loc. cit.*, p. 205. During his famous march to Turin in 1706, which was carried out without magazines, Prince Eugene was actually compelled to besiege and capture a town in order to get rid of part at least of his baggage; Eugene to the Prince of Hesse-Kassel, 5 August 1706, in P. Pieri ed., *Principe Eugenio di Savioa, la Campagna d'Italia del 1706* (Rome, 1936) p. 179.

88 Cf. Vault, *op. cit.*, v, 790–4.

89 See on this point Aubry, *op. cit.*, p. 16.

90 E.g. by Dupré d'Aulnay.

91 Cf. Meixner, *op. cit.*, i, i, 11–14.

Chapter 2

1 It has even been claimed that Napoleon's plan for the 1796 campaign derived directly from Bourcet.

2 It is significant that Vauban, who had a vested interest in the construction of as many fortresses as possible, does not include the ability to stop supply convoys among their many virtues listed in his book.

3 Vauban, *op. cit.*, i, 8–9.

4 Napoléon to Eugène Beauharnais, 16 March 1809, *Correspondance de Napoléon Premier* (Paris, 1863–), xviii, No. 14909.

5 C. J. F. T. Montholon ed., *Recits de la captivité de l'empereur Napoléon à Sainte Hélène* (Paris, 1847) ii, 133–34.

6 'Notes on the town of Erfurt', August 1811, in: Picard ed., *Napoléon, Précepts et Jugements* (Paris, 1913) p. 198; Grouard ed., *Mémoirs écrits en Sainte Hélène* (Paris, 1822–) i, 285.

7 From Saint Helena, he wrote that while he recognized the need for administrators and even made generals of some of them, 'they are repugnant to me'. Montholon, *op. cit.*, i, 452–3.

8 Département de la Guerre/Bureau du Mouvement directive, 25 August 1805, printed in P. C. Alombert and J. Colin, *La Campagne de 1805 en Allemagne* (Paris, 1902) i, 272–5.

9 Berthier to Davout, 27 August 1805; Berthier to Marmont, 28 August 1805; Berthier to Bernadotte, 28 August 1805; *ibid*, 344, 367–8, 369–70.

10 Desportes to Berthier, 30 September 1805; Barbe-Marbois to Dejean, 11 December 1805; *ibid*, ii, 93–4, 94–5.

11 J. Fortescue ed., *Notebook of Captain Coignet* (London, 1928) p. 117.
12 Napoleon to Dejean, 23 August 1805; Napoleon to the Elector of Bavaria, 25 August 1805; *Correspondance*, x, 123, 138–9.
13 List printed in Alombert-Colin, *op. cit.*, i, 583.
14 Napoleon to Berthier, 15 September 1805, *Correspondance*, x, 203–4; Otto to Berthier, 21 September 1805; Murat to Napoleon, 21 September 1805; in Alombert-Colin, *op. cit.*, i, 575, 576–8.
15 Apart from the artillery, Mack had 3,938 horses drawing four-horse wagons for an army numbering 60,000 men. See A. Krauss, *Der Feldzug von Ulm* (Vienna, 1912) pp. 502–3.
16 Soult to Murat, 22 September 1805; Davout to Petiet, 23 September 1805; Alombert-Colin, *op. cit.*, ii, 104, 553–4.
17 Napoleon to Murat, 21 September 1805, *Correspondance*, x, 232–3.
18 Ney order, 26 September 1805, Alombert-Colin, ii, 466–9.
19 Bernadotte to Berthier, 2 October 1805. Depôt de Guerre file No. C² 4.
20 The figures, however, appear less surprising if one reflects that the harvest had just been gathered and that a full year's supply – probably around 3.5 q per head of the population – must have been available in the country. Assuming Pfhul to have been self-sustaining, 2,100 q grain must have been in store there, of which Lannes during his five days' stay cannot have consumed more than 600. The main problem would thus have consisted not so much of finding grain as of grinding it into flour; which explains why mills always formed the first target of marauding parties and had to be put under guard. On food consumption per year, see Perjes, *loc. cit.*, p. 6.
21 Andreossy to Petiet, 4 October 1805, Alombert-Colin, *op. cit.*, ii, 774.
22 Salligny to Soult's divisional commanders, 19 October 1805, *ibid*, iii, 960–1.
23 The total length of the pipeline was about 200 miles, so that each cart would have to travel just over twenty miles per day.
24 Note of 1. Division, 6. corps, 29 September 1805, Depôt de Guerre file C² 3.
25 No precise details about Napoleon's operational plans are known. It seems, however, that he aimed simply at beating the Russians wherever he could find them, and that he expected this to happen early in November before Vienna was reached.
26 Even a hundred years earlier, Austria as a whole was regarded as sufficiently rich to support an army; Perjes, *loc. cit.*, p. 4.
27 Journal Division Friant, in Alombert-Colin, *op. cit.*, iv, 589–90. On 30 October Vandamme's division had a surplus of provisions and was told to send it to a magazine at Riedau; Salligny to Vandamme, 30 October 1805, Depôt de Guerre, file No. C² 6.
28 Murat to Napoleon, 7 November 1805, Alombert-Colin, *op. cit.*, 580–581.
29 Cäncrin, *Über die Militärökonomie im Frieden und Krieg* (St Petersburg, 1821) pp. 230–2.
30 E.g. Mortier to Berthier, 26 November 1805, Depôt de Guerre file C² 8.
31 M. Dumas, *Précis des Evénements Militaires 1799–1814* (Paris, 1822) xiv, 128.

32 On these campaigns see H. de Nanteuil, *Daru et l'Administration militaire sous la Révolution et l'Empire* (Paris, 1966) p. 141ff.

33 See J. Tulard, 'La Depôt de la Guerre et la préparation de la Campagne en Russie', *Revue Historique de l'Armée*, 1969, 3, p. 107ff.

34 Napoleon to Davout, 17 April 1811, *Correspondance Militaire de Napoléon I* (Paris, 1895) vii, No. 1282.

35 Napoleon to Lacue, 13 January 1812, *ibid*, No. 1388.

36 'Instruction dicté par sa Majesté le 16.3.1812 sur le Service administratif de la Grande Armée', Depôt de Guerre file C^2 120.

37 Napoleon to Lacue, 4 April 1812, *Correspondance Militaire*, vii, Nos. 1317, 1324.

38 Note to Berthier, 7 April 1812, Depôt de Guerre file C^2 122.

39 Grande Armée, 'Etat abrégé des principaux objets existants dans les cinq Places d'Allemagne occupés par l'Armée a l'Epoque du 1.5.1812', Dêpot de Guerre file C^2 524.

40 Grande Armée, 'Situation des deux Equippages de Siège de la Grand Armée à l'Epoque du 1.5.1812', *ibid.*

41 G. Perjes, 'Die Frage der Verpflegung im Feldzuge Napoleons gegen Russland', *Revue Internationale d'Histoire Militaire*, 1968, p. 205.

42 To Ségur he said that the provisions were to last until a battle was fought round Vilna; then 'victory must do the rest'. P. de Ségur, *Histoire de Napoléon et de la Grande Armée pendant l'année 1812* n.p., n.d) i, 154. To take his 'memoirs' literally, he intended to fight a battle '200 leagues (500 miles) before Borodino, i.e. on the frontier; *Vie politique et militaire de Napoléon, racontée par lui-même* (Brussels, 1844) ii, p. 194.

43 J. Ullmann, *Studie über die Ausrüstung sowie über das Verpflegs- und Nachschubwesen im Feldzug Napoleon I gegen Russland im Jahre 1812* (Vienna, 1891) pp. 44–5.

44 E. Tarlé, *Napoleon's Invasion of Russia 1812* (London, 1942) p. 65.

45 Details of Pfuel's plan are given by Smitt, *op. cit.*, p. 439ff. Pfuel thought Napoleon would cross the Niemen with nine days' supplies and would need nine days more to reach Drissa; he would, however, be able to requisition supplies in an area measuring 2,500 square miles only. To meet the daily consumption of 250,000 rations, each inhabitant of this area would have to deliver fifty rations – an impossible figure.

46 Napoleon to Berthier, 25 June 1812, in L. G. Fabry, *Campagne de Russie 1812* (Paris, 1912) i, 10.

47 In view of the comments made on this aspect of the problem by subsequent writers it is interesting to read Mortier reporting to Napoleon on 12 July the sighting of 6. train battalion, whose vehicles were loaded with rather more than one ton each, moving 'in good order if slowly'. *Ibid*, p. 121.

48 A. de Caulaincourt, *With Napoleon in Russia, the Memoirs of Général de Caulaincourt* (New York, 1935) p. 86.

49 Both Mortier and Lefebvre reported to Napoleon the 'excellent co-operation' they were encountering on the side of the local population; Fabry, *op. cit.*, i, 599, ii, 24, *passim*.

50 Davout to Napoleon, 20 July 1812; Bourdesoulle to Davout, 27 July 1812; *ibid*, ii, 29, 275.

51 Smitt, *op. cit*, p. 153; Bernardi, *Denkwürdigkeiten aus dem Leben des kaiserlich-russischen Generals von der Infanterie Carl Friedrich von Toll* (Leipzig, 1865) i, 317.

52 Christin to Napoleon, 15 July 1812; Mortier to Napoleon, 13 July 1812; Lefebvre to Mortier, 21 July 1812; Murat to Napoleon, 31 July 1812; Fabry, *op. cit.*, ii, 500, 443, 54–5, 366.

53 Napoleon to Eugène, 18 July 1812; Schwartzenberg to Berthier, 19 July 1812; *ibid*, i, 591, 639.

54 Printed in A. Chuquet, *1812, la Guerre de Russie* (Paris, 1912) i, 60–2.

55 During this time, what was left of the *Grande Armée* repeatedly came upon well stocked magazines at Smolensk, Vitebsk and Vilna; on every one of these occasions, however, the starving troops threw themselves on the provisions, wasting a great part of them and making an orderly distribution impossible.

56 Murat to Napoleon, 6 July 1812, Fabry, *op. cit.*, i, 262. In Lithuania, 1812 was an 'incredibly fruitful' year; Cäncrin, *op. cit.*, p. 81.

57 Perjes, *loc. cit.*, p. 221. For Clausewitz's perorations against Pfuel see *The Campaign of 1812 in Russia* (London, 1843) ch. i.

58 Caulaincourt, *op. cit.*, pp. 66–8.

59 Napoleon crossed the Niemen at the head of 301,000 men; at Smolensk, (15 August) 100,000 had fallen out, and at Borodino he had only 160,000 left. Entering Moscow, the army numbered around 100,000, though various detachments continued to join it later.

60 On 25 September, the army in Moscow still had 877 guns and 3,888 other vehicles in the service of the artillery alone; Grande Armée, 'Situation de l'Artillerie Française et Alliée à l'Epoque du 25.9.1812', Depôt de Guerre file C² 524.

61 See A. von Bülow, *Lerhsätze des neueren Krieges oder reine und angewandete Strategie* (Berlin, 1806 ed.) pp. 26–8.

62 See especially the following: J. Colin, *L'Education Militaire de Napoléon* (Paris, 1901) chps. iii–vi; A. Quimby, *The Background to Napoleonic Warfare* (New York, 1957) chp. xiii; and A. M. J. Hyatt, 'The Origins of Napoleonic Warfare; a Survey of Interpretations', *Military Affairs*, 1966, pp. 177–85.

63 Cf. N. Brown, *Strategic Mobility* (London, 1963) p. 214ff. Here we read that 'a full corps needs...two double-lane improved roads perpendicular to the front and a series of lateral roads'. Napoleon could only have said, Amen.

64 In 1810, Napoleon told le Havre in Spain: 'Pass [Suchet] the order to levy a contribution of many millions from Lerida in order to get the means with which to feed, pay, and dress his army off the country. Make him understand that the war in Spain demands such large forces that it is impossible to send him money; war should nourish war.' *Correspondance*, xx, No. 16521.

Chapter 3

1 Between 28 April and 16 May the Archduke Charles marched over 200 miles without rest from Cham to Aspern, arriving in sufficiently good condition to inflict on Napoleon his first serious reverse.

2 A. de Roginat, *Considérations sur l'art de la Guerre* (Paris, 1816) pp. 439–73.

3 C. von Clausewitz, *On War* (London, 1904) ii, 71–2, 86–107.

4 Meixner, *op. cit.*, i, part ii, p. 17ff.

5 *Ibid*, p. 116ff.

6 Wilhelm Rex, *Militärische Schrifte* (Berlin, 1897) ii, p. 146ff.

7 See on this war Schreiber, *Geschichte des Brandenburgischen Train-Battalions Nr. 3* (Berlin, 1903) pp. 60–80.

8 H. M. Hozier, *The Seven Weeks' War* (London and New York, 1871) p. 80.

9 Schreiber, *op. cit.*, pp. 106–7; E. Bondick, *Geschichte des Ostpreussischen Train Battalion Nr. 1* (Berlin, 1903) pp. 31–2.

10 H. von Moltke, *Dienstschriften* (Berlin, 1898) ii, pp. 254–5.

11 *Ibid*, pp. 250–1. This was not without effect on discipline; E. von Fransecky, *Denkwürdigkeiten* (Berlin, 1913) i, 203, 213.

12 H. Kaehne, *Geschichte des Königlich Preussischen Garde-Train-Battalion* (Berlin, 1903) p. 98.

13 H. von François, *Feldverpflegung bei der höheren Kommandobehörden* (Berlin, 1913) pp. 8–9.

14 *Ibid*, p. 30. During the battle of Königgrätz itself an average of only a single round per rifle was fired.

15 Stoffel, *Rapports Militaires 1866–70* (Paris, 1872) pp. 452–3. I was unable to find exact figures about artillery ammunition, but here too consumption was so low as to make any regular resupply unnecessary.

16 Kaehne, *op. cit.*, pp. 95–6.

17 See W. Heine, 'Die Bedeutung der Verkehrswege für Plannung und Ablauf militärischer Operationen', *Wehrkunde*, 1965, pp. 424–5.

18 K. E. von Pönitz, *Die Eisenbahnen und ihre Bedeutung als militärische Operationslinien* (Adorf, 1853) *passim*.

19 Figures taken from Anon, *De l'Emploi des Chemins de Fer en Temps de Guerre* (Paris, 1869) pp. 5–7; also H. L. W., *Die Kriegführung unter Benützung der Eisenbahnen* (Leipzig, 1868) pp. 6–12.

20 D. E. Showalter, 'Railways and Rifles; the Influence of Technological Developments on German Military Thought and Practice, 1815–1865', (University of Minnesota Diss., 1969) p. 43ff.

21 For these efforts see F. Jacqmin, *Les Chemins de fer pendant la guerre de 1870–71* (Paris, 1872) p. 58ff.

22 H. L. W., *op. cit.*, pp. 15–23.

23 Stoffel, *op. cit.*, pp. 15, 22–3.

24 Moltke to general von der Mülbe, 2 July 1866; Same to Same, 6 July 1866; *Dienstschriften*, ii, pp. 241–2, 249.

25 Moltke to Officers Commanding I and II Armies, 5 July 1866, *Dienstschriften*, ii, p. 249; also H. von Moltke, *Militärische Werke* (Berlin, 1911) iv, 231–2.

26 *Dienstschriften*, ii, 344–5.

27 H. L. W. , *op. cit.*, p. 74.

28 *Ibid*, p. 75ff. These facts are confirmed by a French source: M. Niox, *De l'Emploi des Chemins de Fer pour les Mouvements Stratégiques* (Paris, 1873) pp. 82–7.

29 Showalter, *loc. cit.*, p. 69; H. von Boehn, *Generalstabgeschäfte; ein Handbuch für Offiziere aller Waffen* (Potsdam, 1862) p. 305.

30 Moltke, *Militärische Werke*, i, p. 1ff.

31 Wilhelm Rex, *op. cit.*, ii, 254–65.

32 Showalter, *loc. cit.*, pp. 30–1, 39, 81, 87.

33 F. List, 'Deutschlands Eisenbahnsystem im militärischer Beziehung', in: E. von Beckerath and O. Stühler eds., *Schriften, Reden, Briefe* (Berlin, 1929) iii, 261–5.

34 Showalter, *loc. cit.*, pp. 38, 68; at one point this opinion, too, was shared by Moltke.

35 'Moltkes Kriegslehre', in *Militärische Werke*, iv, p. 210.

36 Moltke, *Dienstschriften*, iii, 130–1.

37 The Prussian forces were transported over the following lines:
  A. Berlin–Hanover–Cologne–Bingerbrück–Neunkirchen.
  B. Leipzig–Harburg–Kreiensen–Mosbach.
  C. Berlin–Halle–Kassel–Frankfurt–Mannheim–Homburg.
  D. Dresden/Leipzig–Bebra–Fulda–Kassel.
  E. Posen–Görlitz–Leipzig–Würzburg–Mainz–Landau.
  F. Münster–Düsseldorf–Cologne–Call.
  The south German troops used the following lines:
  1. Augsburg–Ulm–Bruchsal.
  2. Nördlingen–Crailsheim–Meckesheim.
  3. Würzburg–Mösbach–Heidelberg.
  The entire deployment is analysed in G. Lehmann, *Die Mobilmachung von 1870/71* (Berlin, 1905).

38 Moltke to von Stosch (quartermaster general of the Prussian army), 29 July 1870, *Militärische Werke*, iii, 178; C. von der Goltz, 'Eine Etappenerinnerung aus dem Deutsch–Französischen Kriege von 1870/71', *Beiheft zum Militärwochenblatt*, 1886, p. 311ff.

39 Moltke to all Army- and Corps commanders, 28 July 1870, *Militärische Korrespondenz*, iii, 170.

40 Kaehne, *op. cit.*, pp. 121–2.

41 Fransecky, *op. cit.*, ii, 411.

42 Moltke, *Militärische Werke*, iv, 238–43.

43 Details in C. E. Luard, 'Field Railways and their General Application in War', *Journal of the Royal United Services Institute*, 1874, p. 703.

44 See F. E. Whitton, *Moltke* (London, 1921) pp. 194–5.

45 Hille and Meurin, *Geschichte der preussischen Eisenbahntruppen* (Berlin, 1910) i, 3–22.

46 *Ibid*, pp. 26–8.

47 M. Howard, *The Franco-Prussian War* (London, 1961) p. 376.

48 Moltke, *Dienstschriften*, ii, 288–9.

49 *Ibid*, pp. 327–8.

50 H. Budde, *Die französischen Eisenbahnen in deutschen Kriegsbetriebe 1870/71* (Berlin, 1904) p. 281.

51 This description of the railway situation is based on the masterly analysis in Howard, *op. cit.*, pp. 374–8.

52 See Ernouf, *Histoire des Chemins de Fer Français pendant la Guerre Franco-Prussienne* (Paris, 1874) p. 64.

53 E.g. C. C. Shaw, *Supply in Modern War* (London, 1938) pp. 82–90.
54 Moltke, *Militärische Werke*, iv, 287.
55 Kaehne, *op. cit.*, pp. 129–31.
56 Moltke, *Militärische Werke*, iv, 289.
57 For II Army see Kaehne, *op. cit.*, p. 163; that I Army had to live off the country Moltke himself admits.
58 For the difficulties of supplying the army around Metz see W. Engelhardt, 'Rückblicke auf die Verpflegungverhältnisse im Kriege 1870/71', Beiheft 11 zum *Militärwochenblatt*, 1901, p. 509ff. In 1870, Engelhardt was quartermaster to II Army.
59 This entire episode is well described in A. Hold, 'Requisition und Magazinsverpflegung während der Operationen', *Organ der Militär-Wissenschaftlichen Vereine*, 1878, pp. 484–6.
60 Kaehne, *op. cit.*, pp. 147–8.
61 *Ibid*, p. 171; Blumenthal, *Tagebücher* (Stuttgart, 1901) p. 105, entry for 16 September 1870.
62 Moltke, *Militärische Werke*, iv, 295.
63 Engelhardt, *loc. cit.*, p. 518.
64 Moltke, *Militärische Werke*, iv, 303.
65 H. von Molnar, 'Uber Ammunitions-Ausrüstung der Feld Artillerie', *Organ der Militär-Wissenschaftlichen Vereine*, 1879, pp. 591–3.
66 Von François, *op. cit.*, p. 30.
67 Moltke, *Militärische Werke*, iv, 310; von Hesse, 'Die Einfluss der heutigen Verkehrs- und Nachrichtenmittel auf die Kriegsführung', Beiheft zum *Militärwochenblatt*, 1910, pp. 10–11.
68 Shaw, *op. cit.*, p. 86.
69 Kaehne, *op. cit.*, pp. 211–12.
70 Schreiber, *op. cit.*, pp. 294–5. Provisions to correct the above shortcomings were incorporated in the new regulations of 1874.
71 Moltke, *Militärische Werke*, iv, 311.
72 Typical is the view that, on the subject of the railways, 'nobody save the postgraduate student assiduously seeking a topic for a monograph need look further than the opening pages of Howard, *The Franco-Prussian War*'; a claim made by W. McElwee, *The Art of War, Waterloo to Mons* (London, 1974) pp. 332–3.
73 Moltke, *Dienstschriften*, iii, p. 323ff.
74 Budde, *op. cit.*, pp. 321–2.
75 H. L. W., *op. cit.*, p. 2, footnote 2.
76 Quoted in B. Meinke, 'Beiträge zur frühesten Geschichte des Militär-Eisenbahnwesens', *Archiv für Eisenbahnwesen*, 1938, p. 302.
77 Budde, *op. cit.*, pp. 273–93; also O. Layritz, *Mechanical Traction in War* (Newton Abbot, Devon, 1973 reprint) p. 21ff.
78 Engelhardt, *loc. cit.*, p. 159.

Chapter 4

1 The above figures were compiled from the following sources: O. Riebecke *Was brauchte der Weltkrieg?* (Berlin, 1936) pp. 111–12; E.

Ludendorff, *The General Staff and its Problems* (London, 1921) i, 15–17; E. Wrissberg, *Heer und Waffen* (Leipzig, 1922) i, 82–3.

2 On this point see H. Holborn, 'Moltke and Schlieffen; the Prussian-German School', in E. M. Earle ed., *Makers of Modern Strategy* (New York, 1970 reprint) pp. 200–1.

3 The data were taken from F. von Bernhardi, *On War of Today* (London, 1912) i, 143ff, and C. von der Goltz, *The Nation in Arms* (London, 1913) pp. 241–3.

4 Bernhardi, *op. cit.*, p. 146.

5 Cf. E. A. Pratt, *The Rise of Rail-Power in War and Conquest, 1833–1914* (London, 1916) p. 65; and H. von François, *Feldverpflegungsdienst bei der hoheren Kommandobehörden* (Berlin, 1913) i, 100–1. For the problem of the 'critical distance' as such see Anon, 'Die kritische Transportweite im Kriege', *Zeitschrift für Verkehrwissenschaft*, 1955, pp. 119–24.

6 While space does not allow a list of the literature surrounding the Marne campaign the German side at least is well documented in G. Jäscke, 'Zum Problem der Marne-Schlacht von 1914', *Historische Zeitschrift*, 190, pp. 311–48.

7 Nevertheless, see R. Asprey, *The Advance to the Marne* (London, 1962) pp. 166–70; L. H. Addington, *The Blitzkrieg Era and the German General Staff, 1865–1941* (New Brunswick, N.J., 1971), pp. 15–22; and Ph. M. Flammer, 'The Schlieffen Plan and Plan XVII; a Short Critique', *Military Affairs*, 30, p. 211, for short and conflicting discussions.

8 For Schlieffen's political views see his 'Der Krieg in der Gegenwart', in *Gesammelte Schriften* (Berlin, 1913) pp. 11–22.

9 For Schlieffen's relation to Clausewitz see J. L. Wallach, *Das Dogma der Vernichtungsschlacht* (Frankfurt am Main, 1967) Ch. iii.

10 Strategically, however, such a move had something to be said for it, mainly because it would have drastically cut the distance which the invading German Army had to cover. See Asprey, *op. cit.*, p. 9ff.

11 Politically it carried the danger of British intervention; operationally, it assumed that the French Army would remain behind its fortifications along the Franco-German frontier and tamely await its own destruction.

12 H. von Kuhl, *Der deutsche Generalstab in Vorbereitung und Durchführung des Weltkrieges* (Berlin, 1920) p. 165.

13 There were 7,600 trains carrying troops to the right wing; at sixty trains per (double) line, the increase of the number of railways from six to thirteen reduced the time required from twenty-one to ten days. See C. S. Napier, 'Strategic Movement by Rail in 1914', *Journal of the Royal United Services Institute*, 80, p. 78ff.

14 Exactly what parts of Holland were to be overrun is not clear; see J. J. G. van Voorst tot Voorst, 'Over Roermond!', appended to *De Militaire Spectator*, 92, p. 9.

15 In the 1905 version of the Plan Schlieffen spoke of a front 'from Verdun to Dunkirk'. G. Ritter, *The Schlieffen Plan* (London, 1958) p. 144.

16 H. Rochs, *Schlieffen, Ein Lebens- und Charakterbild für das deutsche Volk* (Berlin, 2nd ed., 1921) p. 21.

17 For the supply arrangements of the German Army at this time see Föst, *Die Dienst der Trains im Kriege* (Berlin, 1908).

18 For German military use of railways see Bernhardi, *op. cit.*, i, p. 144ff; also the much more technical, OHL edited *Taschenbuch für den Offizier der Verkehrstruppen* (Berlin, 1913).

19 W. Gröner, *Lebenserinnerungen* (Göttingen, 1957), p. 73.

20 On the Belgian railroad net see Anon, *Der Krieg, Statistisches, Technisches, Wirtschaftliches* (Munich, 1914) p. 160ff. In this connection it is worth noting that the line from Aix la Chapelle through Liège, Louvain and Brussels to Saint-Quentin, which was destined to serve as the Germans' main artery of supply in Belgium, alone included some 20 tunnels.

21 Quoted in Wallach, *op. cit.*, p. 172.

22 Asprey, *op. cit.*, p. 167ff.

23 Bernhardi, *op. cit.*, i, p. 260; von François, *op. cit.*, i, p. 34.

24 Ritter, *op. cit.*, p. 146, quoting the 1905 version of the Plan.

25 Von der Goltz, *op. cit.*, p. 457.

26 Von François *op. cit.*, i, p. 30. On the eve of the war, the Germans expected consumption of small-arms ammunition to be twelve times that of 1870, whereas that of artillery ammunition was to be four times as large. On this basis, it was assumed that the reserves carried inside each corps would have to be replenished only once during the campaign.

27 Gröner, *op. cit.*, p. 73.

28 See on this point K. Justrow, *Feldherr und Kriegstechnik* (Oldenburg, 1933) p. 249.

29 The Gröner Papers (unpublished, microfilm) reel xviii, item No. 168. See also H. Haeussler, *General Wilhelm Gröner and the Imperial German Army* (Madison, Wisc., 1962) pp. 34–5.

30 F. von Cochenhausen, *Heerführer des Weltkrieges* (Berlin, 1921) pp. 26–7.

31 H. von Moltke, *Erinnerungen, Briefe, Dokumente 1877–1916* (Berlin, 1922) pp. 304ff.

32 Ritter, *op. cit.*, p. 147, quoting the 1905 version of the Plan.

33 See on this point von Falkenhausen, *Der grosse Krieg der Jetztzeit* (Berlin, 1909) p. 217ff.

34 A. von Kluck, *Der Marsch auf Paris und die Marneschlacht 1914* (Berlin, 1920) pp. 14–15. For a graphic description of the congestion during the early days of 1. Army's march see H. von Behr, *Bei der fünften Reserve Division im Weltkrieg* (Berlin, 1919) p. 14.

35 The [USA] Army War College, 'Analysis of the Organization and Administration of the Theater of Operations of the German 1. Army in 1914' unpublished analytical study (Washington, D.C., 1931) pp. 11–12.

36 Voorst tot Voorst, *loc. cit.*, p. 14; see also E. Kabisch, *Streitfragen des Weltkrieges 1914–1918* (Stuttgart, 1924) p. 56.

37 Von Tappen, *Bis zur Marne* (Oldenburg, 1920) p. 8; A. von Baumgarten-Crusius, *Deutsche Heeresführung im Marnefeldzug 1914* (Berlin, 1921) p. 15. The question whether it would have been possible to send in more forces at a later stage is discussed below.

38 OHL ed., *Militärgeographische Beschreibung von Nordost Frankreich, Luxemburg, Belgien und dem südlichen Teil der Niederlande und der Nordwestlichen Teil der Schweiz* (Berlin, 1908) p. 82.
39 Ritter, *op. cit.*, p. 143, quoting the 1905 version of the Plan.
40 The Army War College, 'Analysis of the Organization. . .' p. 12.
41 H. von Kuhl and J. von Bergmann, *Movements and Supply of the German First Army during August and September 1914* (Fort Leavenworth, Kan., 1920) p. 180.
42 Kluck, *op. cit.*, p. 28.
43 W. Bloem, *The Advance from Mons* (London, 1930) p. 38 and *passim*.
44 Kuhl-Bergmann, *op. cit.*, pp. 107–11.
45 Bloem, *op. cit.*, p. 102.
46 *Taschenbuch für den Offizier der Verkehrstruppen*, p. 84; see also von François, *op. cit.*, p. 100. The exact figure depended on the number of horses in each formation, which differed from active to reserve corps since the latter were provided with less artillery.
47 Th. Jochim, *Die Operationen und Rückwärtigen Verbindungen der deutsche 1. Armee in der Marneschlacht 1914* (Berlin, 1935) p. 5.
48 See on this point W. Marx, *Die Marne, Deutschlands Schicksal?* (Berlin, 1932) p. 27.
49 Kuhl-Bergmann, *op. cit.*, pp. 45–6.
50 von der Goltz, *op. cit.*, p. 440; Bernhardi, *op. cit.*, p. 260.
51 Kuhl-Bergmann, *op. cit.*, p. 108.
52 Jochim, *op. cit.*, pp. 6–7.
53 M. von Poseck, *Die deutsche Kavallerie 1914 in Belgien und Frankreich* (Berlin, 1921) p. 40.
54 *Ibid*, pp. 20, 27, 29.
55 *Ibid*, p. 30.
56 von Haussen, *Erinnerungen an den Marnefeldzug 1914* (Leipzig, 1920) p. 171; K. von Helfferich, *Der Weltkrieg* (Berlin, 1919) vol. ii, p. 17.
57 The Army War College, 'Analysis of the Organization. . .' p. 18.
58 In August 1914 1. and 2. Armies each had eighteen companies, 3. Army had nine, 4. and 5. Armies five each. A motor transport company consisted of nine trucks with trailers (overall capacity fifty-four tons) and a few other vehicles serving for control, maintenance and repairs.
59 The Army War College, 'Analysis of the Organization. . .' p. 18. In 1941, this particular error was to be repeated.
60 Kuhl–Bergmann, *op. cit.*, p. 196.
61 *Ibid*, p. 52; see also Jochim, *op, cit.*, p. 129.
62 Kluck, *op. cit.*, pp. 64–65.
63 Kuhl–Bergmann, *op. cit.*, pp. 120–1. The shortage of ammunition, as well as the inability of their light infantry (Jäger) element to keep up, greatly reduced the combat-power of the cavalry divisions throughout the campaign.
64 Jochim, *op. cit.*, p. 24.
65 Gröner, *op. cit.*, p. 175.
66 W. Müller-Löbnitz, *Die Sendung des Oberstleutnants Hentsch am 8–10 September 1914* (Berlin, 1922) p. 30, and appendix 1.
67 Jochim, *op. cit.*, pp. 129–31.

68 Cf. Hauessler, *op. cit.*, pp. 61–2.

69 Reichsarchiv ed., *Der Weltkrieg, Das deutsche Feldeisenbahnwesen* (Berlin, 1928) i, appendix 6, pp. 221–2. The installations in question were a tunnel near Homburg, a bridge over the Ourthe at Melreux and a viaduct at Haversin.

70 See on this point Bloem's description, *op. cit.*, of his journey back from Belgium after being wounded on the Aisne.

71 M. Heubes, *Ehrenbuch der deutschen Eisenbahner* (Berlin, 1930) pp. 49–50.

72 Gröner, *op. cit.*, p. 190.

73 Cf. W. Kretschmann, *Die Wiederherstellung der Eisenbahnen auf dem Westlichen Kriegsschauplatz* (Berlin, 1922) p. 36.

74 Jochim, *op. cit.*, p. 27.

75 Justrow, *op. cit.*, p. 250.

76 Kuhl-Bergmann, *op. cit.*, pp. 215–16.

77 Kretschmann, *op. cit.*, p. 64.

78 Statistics about the time required to remove various kinds of obstacles are given in C. S. Napier, 'Strategic Movement over Damaged Railways in 1914', *Journal of the Royal United Services Institute*, 81 pp. 317, 318.

79 See on this point Ritter, *op. cit.*, p. 59ff.

80 See on this point Justrow, *op. cit.*, pp. 244–5.

81 R. Villate, 'L'état matériel des armeés allemandes en Aôut et Septembre 1914', *Revue d'Histoire de la Guerre Mondiale*, 4, pp. 313–25.

82 Kluck, *op. cit.*, p. 90; Haussen, *op. cit.*, pp. 178–9.

83 Baumgarten-Crusius, *op. cit.*, pp. 55, 63–4, 78; H. von François, *Marneschlacht und Tannenberg* (Berlin, 1920) pp. 118–20.

84 A. von Schlieffen, *Cannae* (Berlin, 1936) p. 280.

85 For details see Reichsarchiv ed., *Der Weltkrieg* (Berlin, 1921) i, pp. 139, 152.

86 See especially H. Gackenholtz, *Entscheidung in Lothringen 1914* (Berlin, 1933) and Tappen, *op. cit.*, pp. 14–16.

87 Kuhl–Bergmann, *op. cit.*, p. 216.

88 Justrow, *op. cit.*, p. 250. See also W. Gröner, *Der Feldherr wider Willen* (Berlin, 1931) p. 12ff.

89 H. von Moltke, *Die deutsche Tragödie an der Marne* (Berlin, 1934) p. 19.

90 Based on the 120 trains mentioned above. Since a railway wagon was equivalent to three trucks, each train could only have been replaced by 150 motor vehicles.

91 G. L. Binz, 'Die stärkere Battalione', *Wehrwissenschaftliche Rundschau*, 9, pp. 139–61.

92 The [USA] Army War College, 'Analytical Study of the March of the German 1. Army, August 12–24 1914', (unpublished staff study, Washington D.C., 1931) pp. 7–9. Even as it was, the fact that Kluck's corps had to share their roads when passing between Brussels and Namur made it physically impossible for some of them to be present at the battle of Mons on 24 August.

93 Kuhl–Bergmann, *op. cit.*, pp. 227–33.

94 To cater for a consumption of 250 tons a day six echelons of motor companies, all travelling at 100 miles per day, would have had to

shuttle over 300 miles from the German frontier to the Marne. To achieve the required overall capacity of 1,500 tons, 500 three-ton lorries were necessary. The time required for loading, unloading and repairs is not taken into account by these figures.

95 Tappen, *op. cit.*, pp. 14–15.

96 There was at this time available a daily train, carrying 250–300 tons according to the kind of load, for each of the corps involved.

97 See General Föst, inspector-general of the trains-service, in M. Schwarte ed., *Die militärische Lehre des grossen Krieges* (Berlin, 1923) p. 279.

98 Quoted in Kuhl–Bergmann, *op. cit.*, pp. 180–1.

99 In his introduction to Ritter, *op. cit.*, pp. 6–7.

100 See A. M. Henniker, *Transportation on the Western Front, 1914–18* (London, 1937) p. 103.

Chapter 5

1 H. Rohde, *Das deutsche Wehrmachttransportwesen im Zweiten Welt-krieg* (Stuttgart, 1971) pp. 174–75; E. Kreidler, *Die Eisenbahnen im Machtbereich der Assenmächte wahrend des Zweiten Weltkrieges* (Göttingen, 1975) p. 22; the relationship between roads and railroads is well discussed by R. J. Overy, 'Transportation and Rearmament in the Third Reich', *The Historical Journal*, 1973, pp. 391–93.

2 For some figures see A. G. Ploetz, *Geschichte des Zweiten Weltkrieg* (Würzburg, 1960) ii, 687.

3 F. Friedenburg, 'Kan der Treibstoffbedarf der heutigen Kriegsführung überhaupt befriedigt werden?', *Der deutsche Volkswirt*, 16 April 1937; memorandum by General Thomas, 24 May 1939, in *International Military Tribunal* ed., *Trials of the Major War Criminals* (Munich, 1946–) doc. No. 028–EC, esp. pp. 124, 130.

4 Windisch, *Die deutsche Nachschubtruppe im Zweiten Weltkrieg* (Munich, 1953) pp. 38–9.

5 H. A. Jacobsen, 'Motorisierungsprobleme im Winter 1939/40', *Wehrwissenschaftliche Rundschau*, 1956, p. 513.

6 For the entire organization, see I. Krumpelt, *Das Material und die Kriegführung* (Frankfurt am Main, 1968) p. 108ff.

7 Rohde, *op. cit.*, p. 212; also R. Steiger, *Panzertaktik* (Freiburg i.B., 1973) pp. 146, 155.

8 F. Halder, *Kriegstagebuch* (Stuttgart, 1962, henceforward *KTB*/Halder) i, pp. 179–82, entries for 3, 4 February 1940.

9 H. A. Jacobsen, *Fall Gelb* (Wiesbaden, 1957) p. 130.

10 E. Wagner, *Der Generalquartiermeister* (Munich and Vienna, 1963) pp. 256–8; Krumpelt, *op. cit.*, p. 130ff.

11 Wagner, *op. cit.*, p. 184.

12 See my *Hitler's Strategy 1940–1941; the Balkan Clue* (Cambridge, 1973) p. 221, footnote No. 130.

13 *KTB*/Halder, ii, 256–61, 420, 421–2, entries for 28 January, 19, 20 May 1941; H. Greiner, 'Operation Barbarossa', unpublished study No. C–0651 at the Imperial War Museum, London, pp. 60–1. On the problem of raw materials as a whole see G. Thomas, *Geschichte der*

deutschen Wehr- und Rüstungswirtschaft 1918–1943/45 (Boppard am Rhein, 1966) appendix 21, esp. pp. 530ff.

14 *KTB*/Halder, 422, entry for 20 May 1941.

15 Krumpelt, *op. cit.*, p. 187.

16 Cf. R. Cecil, *Hitler's Decision to Invade Russia 1941* (London, 1975) pp. 128–9.

17 *KTB*/Halder, ii, 384, entry for 29 April 1941.

18 Windisch, *op. cit.*, pp. 41–2.

19 Based on the lorries being driven ten hours per day at an average speed of 12 miles an hour. It does not, however, make allowance for the fact that twenty to thirty-five per cent of all the vehicles would be under repair at any given time.

20 Equal to 360 cubic meters, carried by three companies of four vehicle-columns each.

21 *KTB*/Halder, ii, 414, entry for 15 May 1941.

22 Windisch, *op. cit.*, pp. 22, 41–2.

23 Cf. M. Bork, 'Das deutsche Wehrmachttransportwesen – eine Vorstufe europäischer Verkehrsführung', *Wehrwissenschaftliche Rundschau*, 1952, p. 52.

24 'Vortrag des Herrn O. B. der Eisenbahntruppen', 11 June 1941, German Military Records (microfilm, henceforward GMR/T–78/259/6204884ff.

25 H. Pottgiesser, *Die deutsche Reichsbahn im Ostfeldzug 1939–1944* (Neckargemund, 1960) pp. 24–5.

26 B. Mueller-Hillebrand, *Das Heer* (Frankfurt am Main, 1956) ii, p. 81ff.

27 A curious result of this duality was the different estimates of requirements by the two authorities. Whereas Wagner reckoned in tons, Gercke had to count each train as a train regardless of its load; hence they could, and did, arrive at very different results when called upon to decide whether requirements could be, or were, met.

28 See also Krumpelt, *op. cit.*, pp. 151–2.

29 Tagesmeldung der Genst.d.H/Op.Abt., 24 June 1941, printed in *Kriegstagebuch des Oberkommando der Wehrmacht* (Frankfurt am Main, 1965, henceforward *KTB*/OKW) i, p. 493.

30 *KTB*/Halder, iii, 62–3, entry for 11 July 1941.

31 Mueller-Hillebrand, *op. cit.*, ii, 81.

32 *KTB*/Halder, iii, 32, entry for 1 July 1941.

33 *Ibid*, 170, entry for 11 August 1941.

34 H. Teske, *Die Silbernen Spiegel* (Heidelberg, 1952) p. 131.

35 *KTB*/Genst.d.H/Op.Abt., 5, 13 July 1941, in *KTB*/OKW, i, 427, 433.

36 Vortrag des Oberst Dybilasz, 5 January 1942, p. 15, GMR/T–78/259/6204741.

37 Rohde, *op. cit.*, p. 173. Another reason for the limited capacity of the Russian railroads was the quality of the personnel, which had been supplied by the *Reichsbahn* and which Halder regarded as insufficiently flexible and too slow.

38 Teske, *op. cit.*, p. 132.

39 W. Haupt, *Heeresgruppe Nord 1941–1945* (Bad Nauheim, 1966) p. 22.

40 Having completed this task, the German troops were to move against Leningrad. The order of priorities, however, was not entirely clear; see

on this point A. Seaton, *The Russo-German War* (London, 1971) p. 105.

41 KTB/Pz.Gr. 4/0.Qu, 24, 27 June 1941, *Militärgeschichtliches Forschungsamt* (MGFA), Freiburg, file No. 22392/1.

42 *Ibid*, entry for 1 July 1941.

43 *Ibid*, entries for 9, 10 July 1941.

44 *Ibid*, entry for 11 July 1941.

45 *Ibid, ibid*; 'Vortrag beim Chef [der Eisenbahntruppen]', 10 July 1941, GMR/T–78/259/6204892; *KTB*/Halder, iii, 34, entry for 2 July 1941.

46 KTB/Aussenstelle OKH/Gen.Qu/HGr Nord, 30 June 1941, *ibid*, T–311/111/7149931.

47 *Ibid*, 3 July 1941, *ibid*, 7149920.

48 *KTB*/Halder, iii, 148–9, entry for 3 August 1941.

49 KTB/Aussenstelle OKH/Gen.Qu/HGr Nord, entry for 24 July 1941, GMR/T–311/111/714889.

50 *Ibid, ibid.*

51 *Ibid*, entries for 2, 9 August, 9 September 1941, *ibid*, 7149876–7, 7149867, 7149831.

52 KTB/Pz.Gr. 4/0.Qu, entries for 17, 18 July 1941, MGFA file 22392/1.

53 KTB/Aussenstelle OKH/HGr Nord, 17 July, 1941, GMR/T–311/111/7149931.

54 *KTB*/Halder, iii, 129, 133, entries for 29, 31 July 1941.

55 KTB/Aussenstelle OKH/Gen.Qu/HGr Nord, 22 July 1941, GMR/T–311/111/7149902.

56 *Ibid*, 28, 6, 13 August 1941, *ibid*, 7149923, 7149863; KTB/Pz.Gr. 4/0.Qu, entry for 1 August 1941, MGFA file 22392/1.

57 *Ibid*, 25 July 1941.

58 *Ibid*, 31 July 1941.

59 Seaton, *op. cit.*, p. 108ff.

60 *KTB*/Halder, iii, 124, entry for 26 July 1941.

61 KTB/Pz.Gr. 4/0.Qu, entry for 21 July 1941, MGFA file 22392/1.

62 Haupt, *op. cit.*, p. 62; W. Chales de Beaulieu, *Generaloberst Erich Hoepner* (Neckargemund, 1969) p. 158ff.

63 KTB/Pz.Gr. 4/0.Qu, entry for 2 August 1941, MGFA file 22392/1.

64 Haupt, *op. cit.*, pp. 69, 88–9.

65 KTB/Pz.Gr. 4/0.Qu. entries for 8–30 August 1941, *passim*, MGFA file 22392/1.

66 'Directive No. 21', 18 December 1940, printed in H. R. Trevor-Roper, *Hitler's War Directives* (London, 1964) pp. 50–1; OKH/Genst.d.H/ Aufmarschanweisung Barbarossa, 31 January 1941, printed in *KTB*/ Halder, ii, pp. 463–9.

67 KTB/Pz.Gr. 1/0.Qu, entries for 14, 16, 25 August 1941, MGFA file 16910/46.

68 *Ibid*, entry for 1 July 1941.

69 *Ibid*, entry for 20 July 1941; *KTB*/Halder, iii, 94, entry for 19 July 1941.

70 KTB/Pz.Gr. 1/0.Qu, entry for 20 July 1941, MGFA file 16910/46; also Teske, *op, cit.*, pp. 120. 127.

71 Vortrag ObdH und Chef Genst.d.H beim Führer, 8 July 1941, *KTB*/ OKW, i, 1021; *KTB*/Halder, iii, 108, entry for 23 July 1941; also Seaton, *op. cit.*, p. 141.

72 *KTB*/Halder, iii, 138–9, entry for 1 August 1941.

73 KTB/Pz.Gr. 1/0.Qu, 22, 23, 24 August 1941, MGFA file 16910/46.

74 *Ibid*, 20 August, 1 September 1941.

75 KTB/Aussenstelle OKH/Gen.Qu/HGr Sud, 16 August–30 September 1941, p.4, MGFA file 27927/1. This particular diary is written in the form of a narrative, hence the absence of a date.

76 Aussenstelle OKH/Gen.Qu/HGr Sud, 'Besondere Anordnungen für die Versorgung No. 105', 6 September 1941, GMR/T–311/264/000071–2.

77 KTB/Aussenstelle OKH/Gen.Qu/HGr Sud, p. 8, MGFA file 27927/1.

78 KTB/Pz.Gr. 1/0.Qu, entry for 3 October 1941, MGFA file 16910/46.

79 *Ibid*, entries for 17, 20 October 1941.

80 *Ibid*, 24 October 1941.

81 KTB/Aussenstelle OKH/Gen.Qu/HGr Sud, October 1941, p. 4, MGFA file 27927/1.

82 *Ibid*, p. 8. There is a map of the railways available to Army Group South at this time in Bes. Anlagen No. 7 to KTB/Aussenstelle OKH/Gen.Qu/HGr Sud, MGFA file 27928/8.

83 *KTB*/Halder, iii, 278, 279, entries for 3, 4 November 1941.

84 KTB/Aussenstelle OKH/Gen.Qu/HGr Sud, November 1941, pp. 3–8, MGFA file 27927/1.

85 Aussenstelle OKH/Gen.Qu/HGr Sud, 'Vorschlag zur Lösung des Brennstoffproblems', 27 November 1941; 'Versorgung des Weharmachtsbefehlhaber Ukraine', 8 November 1941, GMR/T–311/264–000408–11, 000441–43.

86 KTB/Pz.Gr. 1/0.Qu. 1 December 1941, MGFA file 169/46.

87 No very clear idea existed as to what was to happen after the capture of Smolensk. In 'Directive No. 21' Hitler had expressed his intention of going over to the defensive in this sector, while sending the *Panzergruppen* to the left and the right in support of the neighbouring army groups; however, OKH was dead against this plan and was quietly hoping to sabotage it.

88 KTB/Pz.Gr. 2/0.Qu, 23, 25 June 1941, MGFA file RH/21–2/v 819.

89 *Ibid*, 30 June 1941; KTB/AOK 9/0.Qu, entries for 22 June to 6 July 1941 *passim*, MGFA file 139041/1.

90 KTB/Pz.Gr. 2/0.Qu, 2 July 1941, MGFA file RH/21–2/v 819.

91 *Ibid*, 10, 22 July 1941. See also Krumpelt, *op. cit.*, pp. 165–7, where the entire idea of building a pocket at Smolensk is criticized from a logistic point of view.

92 *KTB*/Halder, iii, 32, 66, entries for 1, 11 July 1941; AOK 9/0.Qu, Abendmeldung v. 7 July 1941, MGFA file 13904/4.

93 AOK 9/0.Qu. Abendmeldung v. 10, 11 July 1941, *ibid*.

94 *KTB*/Halder, iii, 71, 78, entries for 13, 14 July 1941. On 17 July, this estimate was revised further downwards.

95 AOK 9/0.Qu to OKH/Gen.Qu, 19 July 1941, MGFA file 13904/4; AOK 2/0.Qu Tagesmeldungen, 22 June–10 July 1941 *passim*, MGFA file 16773/14.

96 AOK 9/0.Qu to OKH/Gen.Qu, 4 August 1941, MGFA file 13904/4; AOK 2/0.Qu, Tagesmeldung v, 12 August 1941, MGFA file 16773/14.

97 AOK 9/0.Qu, Tagesmeldungen v. 14, 15, 18 August 1941, MGFA file 13904/4.

98 Pz.Gr. 2/0.Qu, Tagesmeldung v. 24 August 1941, MGFA file RH/21–2/ v 829; 'Besprechung gelegentlich Anwesenheit des Führers und Obersten Befehlhaber der Wehrmacht bei Heeresgruppe Mitte am 4 August 1941', No. 31, file AL 1439 at the Imperial War Museum, London.

99 AOK 9/0.Qu, Tagesmeldungen v. 21, 23, 31 August 1941, MGFA file 13904/4; AOK 9/0.Qu/IVa, 'Täteskeitsbericht für die Zeit 17–23 August 1941', MGFA file 13904/1; also I. Krumpelt, 'Die Bedeutung des Transportwesens für den Schlachterfolg', *Wehrkunde*, 1965, pp. 466–7.

100 Seaton, *op, cit.*, p. 147.

101 AOK 2/0.Qu, Tagesmeldungen v. 3–15 September 1941 *passim*, MGFA file 16773/14.

102 *KTB*/Halder, iii, 178–9, 181, entries for 15, 16 August 1941.

103 *Ibid*, 120, 178–80, 245, entries for 26 July, 15 August, 22 September 1941.

104 *Ibid*, 196, entry for 25 August 1941.

105 AOK 9/0.Qu to OKH/Gen.Qu, 14 September 1941, MGFA file 13904/4.

106 AOK 4/0.Qu No. 1859/41 g. v. 13 September 1941, 'Versorgungslage der Armee', Anlagen zum KTB, MGFA file 17847/3.

107 KTB/AOK 9/0.Qu, 30 September 1941, MGFA file 13904/2.

108 Cf. footnote 106 *supra*.

109 KTB/Halder, iii, 242, 245, 252, entries for 12, 22, 26 September 1941.

110 Pz.Gr. 4/0.Qu, Tagesmeldung v. 4 November 1941, Anlagen zum KTB, MGFA file 13094/5.

111 AOK 4/0.Qu, Abendmeldung v. 8 October 1941, Anlagen zum KTB, MGFA file 17847/4.

112 Seaton, *op. cit.*, p. 190: 'with another three weeks' dry, mild and clear weather, [Army Group Center] would inevitably have been in Moscow.'

113 Pz.Gr. 2/0.Qu, Tagesmeldungen v. 11 October, 1, 9, 13, 23 November 1941, Anlagen zum KTB, MGFA file RH/21–2/v829; see also H. Guderian, *Panzer Leader* (London, 1953) pp. 180–9.

114 AOK 9/0.Qu to OKH/Gen.Qu, 13 November 1941, MGFA file 13094/5.

115 Pz.Gr. 4/0.Qu, Tagesmeldungen v. 18, 22, 25 October 1941, Anlagen zum KTB, MGFA file 22392/22. For this formation see also Chales de Beaulieu, *op. cit.*, pp. 209–10.

116 AOK 2/0.Qu, Tagesmeldungen v. 21. 24, 26–31 October 1941, Anlagen zum KTB, MGFA file 16773/14.

117 AOK 4/0.Qu, Abendmeldungen v. 20–28 October 1941, Anlagen zum KTB, MGFA file 17847/4. At the end of this period, the troops had eighty per cent of a basic ammunition load, one to three and a half loads of fuel, and two to three loads of food.

118 *Ibid*, Abendmeldungen v. 6–10 November 1941. During this period stocks with the troops continued to rise.

119 AOK 4/0.Qu, No. 406/41, 'Beurteiling der Versorgungslage am 13 November 1941', Anlagen zum KTB, MGFA file 17847/4.

120 Aussenstelle OKH/Gen.Qu/HGr Sud, No. 1819/41 g., 26 October 1941, appendix 5, GMR/T–311/264/000466.
121 Pottgiesser, *op. cit.*, p. 35.
122 AOK 2/0.Qu, Tagesmeldungen, 17 November–2 December 1941, Anlagen zum KTB, MGFA file 16773/14.
123 AOK 9/0.Qu, Tagesmeldungen, 9–23 November 1941, Anlagen zum KTB, MGFA file 13904/5.
124 Pz.Gr. 4/0.Qu, Tagesmeldungen, 18 November–6 December 1941, Anlagen zum KTB, MGFA files 22392/22, 22392/23. For *Panzergruppe* 3 see Chales de Beaulieu, 'Sturm bis Moskaus Tore', *Wehrwissenschaftliche Rundschau*, 1956, pp. 360–4.
125 OAK 4/0.Qu, Abendsmeldungen, 22–30 November 1941, Anlagen zum KTB, MGFA file 17847/5.
126 Eckstein memoirs, printed in Wagner, *op. cit.*, pp. 288–9.
127 *KTB*/Halder, iii, 111, entry for 25 July 1941.
128 A good example for winter preparations is Aussenstelle OKH/Gen.Qu/ HGr.Sud, No. 181941 g., 26 October 1941, 'Anordnungen für die Versorgung im Winter 1941–42/, GMR/T–311/264/000446–73; in it, are listed no less than 73 separate orders, the earliest of which dates to 4 August, dealing with every conceivable aspect of the problem.
129 Wagner, *op. cit.*, pp. 313–17.
130 Windisch, *op. cit.*, pp. 41–2.
131 *KTB*/Halder, iii, 88, entry for 15 July 1941.
132 B. H. Liddell Hart, *History of the Second World War* (London, 1973) pp. 163–5, 177.
133 *KTB*/Halder, iii, 292–3, entry for 17 November 1941.
134 So short was manpower that orders to this effect had to be given even before the start of the campaign; KTB/AOK 9/0.Qu, 'Besprechung Gen.Qu. in Posen, 9–10 June 1941', p. 2, MGFA file 13904/1.
135 Grukodeis Nord/Ia, 'Einsatzbefehl No. 11' 28 August 1941, GMR/ T–78/259/6205110.
136 *KTB*/Halder, iii, 149, entry for 3 August 1941.
137 AOK 4/Ia, No. 3944/41 v. 21 November 1941, Anlagen zum KTB, MGFA file 17847/5.
138 Gen.Qu/Qu. 2, 'Erfahrungen aus dem Ostfeldzug über die Versorgungsführung', 24 March 1942, p. 4, MGFA file H/10–51/2.
139 *Ibid*, p. 5.
140 OKH/Genst.d.H/Org.Abt. (III), 'Notizien für KTB', 11 August 1941, GMR/T–78/414/6382358–59; also KTB/AOK 9/0.Qu, 'Planbesprechung 19–21.5.1941', p. 6, MGFA file 13904/1.
141 See on this point my 'Warlord Hitler; Some points Reconsidered', *European Studies Review*, 1974, p. 76.
142 OKH/Gen.Qu, No. I 23 637/41, 12 August 1941, MGFA file RH/21–2/ v 823.
143 Mueller-Hillebrand, *op. cit.*, ii, 18ff.
144 Seaton, *op. cit.*, p. 222.
145 See A. Hillgruber, *Hitlers Strategie* (Frankfurt am Main, 1965) pp. 533ff.

Chapter 6

1 E.g., K. Assmann, *Deutsche Schicksahljahre* (Wiesbaden, 1951) p. 211, and D. Young, *Rommel* (London, 1955) pp. 201–2.
2 Hillgruber, *op. cit.*, pp. 190–2; L. Gruchman, 'Die "verpassten strategischen chancen" der Assenmächte im Mittelmeerraum 1940–41', *Vierteljahrshefte für Zeitgeschichte*, 1970, pp. 456–75.
3 B. H. Liddell Hart ed, *The Rommel Papers* (New York, 1953) pp. 199–200; E. von Rintelen, 'Operation und Nachschub', *Wehrwissenschaftliche Rundschau*, 1951, 9–10, pp. 46–51.
4 The best treatment is still D. S. Detwiller, *Hitler, Franco und Gibraltar* (Wiesbaden, 1962).
5 This was Hitler's own view; *KTB*/Halder, ii, 164–65, entries for 4, 24.11.1940.
6 For details see E. D. Brant, *Railways of North Africa* (Newton Abbot, Devon, 1971) pp. 180–1.
7 W. Stark, 'The German Afrika Corps', *Military Review*, July 1965, p. 97.
8 B. Mueller-Hillebrand, 'Germany and her Allies in World War II', (unpublished study, U.S. Army Historical Division MS No. P–108) part I, pp. 82–3, GMR/63–227.
9 *The Rommel Papers*, p. 97.
10 Hitler–Mussolini conversation, 4 October 1940, *Documents on German Foreign Policy* (Washington and London, 1948–, henceforward *DGFP*) series D. vol. xi, No. 159.
11 B. H. Liddell Hart, *The German Generals Talk* (London, 1964) pp. 155–6.
12 *KTB*/OKW, i, 253, entry for 9 January 1941.
13 *Ibid*, i, 301, entry for 3 February 1941.
14 *Ibid*, i, 292–3, entry for 1 February 1941.
15 This was 200 miles; *KTB*/Halder, iii, 106, entry for 23 July 1941.
16 OKH/Genst.d.H/Gen.Qu No. 074/41 g.Kdos, 11 February 1941, 'Vortragnotiz über Auswirkungen des Unternehmen Sonneblume auf das Unternehmen Barbarossa', GMR/T–78/324/6279177–79.
17 *KTB*/OKW, i, 318, entry for 11 February 1941.
18 *KTB*/Halder, ii, 259, entry for 28 January 1941; OKH/Genst.d.H/Gen.Qu to OKW/WFSt/Abt.L, No. I/0117/41 g.Kdos, 31 March 1941, GMR/T–78/6278948–49; OKW/WFSt/Abt.L (I.Op.) No. 4444/41 g.Kdos, 3 April 1941, printed in *KTB*/OKW, i, 1009.
19 Cf. L. H. Addington, *The Blitzkrieg Era* (New Brunswick, N.J., 1971) p. 163.
20 DAK/Ia to OKH/Genst.d.H/Op.Abt, No. 63/41 g. Kdos, 9 March 1941, GMR/T–78/324/6278950; *KTB*/Halder, ii, 451, entry for 11 June 1941.
21 OKH/Genst.d.H/Gen.Qu IV, No. 170/41 g.Kdos, 27 May 1941, GMR/T–78/324/6278954.
22 Representative of foreign ministry with German armistice commission to foreign ministry, 28 April 1941, *DGFP*, D. xii, No. 417. No formal agreement on the matter, however, was reached until June, and then the French put so many obstructions in the way that in the end hardly

any *materiel* reached the Germans. Cf. M. Weygand, *Recalled to Service* (London, 1952) pp. 337–43.

23 Hitler–Darlan conversation, 11 May 1941, *DGFP*, D. xii, No. 491; protocols signed at Paris on 27 and 28 May 1941, *ibid*, No. 559, part ii, 'agreement with regard to North Africa'.

24 Cf. E. Jäckel, *Frankreich in Hitlers Europa* (Stuttgart, 1966) pp. 171–9.

25 OKH/Genst.d.m/Op.Abt. (IIb) No. 35512/41 g.Kdos, 'Tagesmeldung von 10.5.1941', GMR/T–78/307/6257983; DAK to OKH/Genst.d.H/ Op.Abt, No. 419 (Abendmeldung von 16.5.1941), *ibid.*, 324/6279335–36.

26 M. A. Bragadin, *The Italian Navy in World War II* (Annapolis, Md., 1957) p. 72.

27 Addington, *op, cit.*, p. 163.

28 W. Baum and E. Weichold, *Der Krieg der Assenmächte im Mittelmeer-Raum* (Göttingen, 1973) p. 134.

29 G. Rochat, 'Mussolini Chef de Guerre', *Revue d'Histoire de la deuxième guerre mondiale*, 1975, pp. 62–4.

30 Chef OKW to foreign minister, 15 June 1941, *DGFP*, D, xii, No. 633.

31 *KTB/OKW*, i, 394, entry for 11 May 1941.

32 Early in May, for instance, the port was blocked; OKH/Genst.d.H/ Op.Abt (IIb), No. 35461/41 g.Kdos, 2 May 1941, GMR/T–78/307/ 6285972.

33 M. Gabriele, 'La Guerre des Convois entre l'Italie et l'Afrique du Nord', in: *La Guerre en Mediterranée, 1939–1945* (ed. Comité d'Histoire de la Deuxième Guerre Mondiale, Paris, 1971) p. 284.

34 *KTB*/Halder, ii, 377, entry for 23 April 1941.

35 OKH/Genst.d.H/Op.Abt to Deutsche General beim Hauptquartier der italienische Wehrmacht, No. 1633/41 g.Kdos, 8 June 1941, GMR/ T–78/324/6279035; and DAK/Ia to OKH/Genst.d.H/Op.Abt, No. 414/41 g.Kdos, *ibid*, 6279151–52.

36 OKH/Genst.d.H/Op. Abt to OKW/WFSt/Abt.L, No. 1380/41 g.Kdos, 21 July 1941, *ibid*, 6278960.

37 OKH/Genst.d.H/Op.Abt to DAK, No. 1299/41 g.Kdos, 28 June 1941, *ibid*, 6279170–72; and OKH/Genst.d.H/Op.Abt No. 1292/41 g.Kdos, 3 July 1941, *ibid*, 6279145.

38 SKL/Op.Abt to OKH/Genst.d.H/Op.Abt, No. 1509/41 g.Kdos, 12 September 1941, GMR/T–78/324/6278993. General Westphal's (*Erinnerungen*, Mainz, 1975, p. 127) claim that nobody had troubled to inform Rommel of this is therefore false.

39 DAK/Gen.Qu 1 to OKH/Genst.d.H/Gen.Qu, No. 41/41 g.Kdos, 13 July 1941, GMR/T–78/324/6279249–51.

40 Gabriele, *loc. cit.*, p. 292.

41 DAK/Ia to OKH/Genst.d.H/Op.Abt, No. 48/41 g.Kdos, 25 July 1941, GMR/7–78/324/6279246–47.

42 SKL/Ia to ObdH/Genst.d.H/Op.Abt, No. 1321/41 g.Kdos, 19 August 1941, GMR/T–314/15/6298989–90.

43 OKH/Genst.d.H/Op.Abt, No. 1496/41 g.Kdos, 12 September 1941, 'Notiz zu Der dtsch. Gen. b. H.Q. d. Ital. Wehrmacht', No. 2448/41 g.Kdos, 6 September 1941, *ibid*, T–78/324/6279017–20. On 20 September the Luftwaffe in Greece was told to detail units to protect 'the most

important' convoys to Africa, but the definition of what was 'most important' was left to the local commander.

44 Cf. the Morgenmeldungen of OKH/Genst.d.H/Op.Abt for 15, 22 October, 5 November 1941, *ibid*, 307/6258236ff.

45 Der dtsch. Gen. b. H.Q. d. Ital. Wehrmacht to OKH/Genst.d.H/Op. Abt, Nos. 15066/41 and 15088/41 g.Kdos, 30.10.1941, *ibid*, 324/6279032, 6278991–92; 'Verbale del Colloquio tra l'Eccelenza Cavallero ed il Maresciallo Keitel, 25.8.1941', p. 5, Italian Military Records (IMR)/T–821/9/000326.

46 Der dtsch. Gen. b. H.Q. d. Ital. Wehrmacht to OKH/Genst.d.H/Op. Abt., No. 15080/41 g.Kdos, 9 October 1941, GMR/T–78/324/6279023–24.

47 Pz. AOK Afrika to OKH/Genst.d.H/Op.Abt No. 39/41, 12 September 1941, *ibid*, 6278997–99; and OKW/WFSt/Abt.L to ObdH/Genst.d.H/Op.Abt, No. 441587/41 g.Kdos, 26 September 1941, *ibid*, 6279001–2.

48 S.O Playfair, *The Mediterranean and the Middle East* (London, 1956) ii, 281.

49 Pz.AOK Afrika/O.Qu No. 285/41 g.Kdos, 12 September 1941, GMR/T–314/15/000992.

50 Der dtsch. Gen. b. H.Q. d. Ital. Wehrmacht to OKH/Genst.d.H/Op. Abt, No. 150104/41 g.Kdos, 11 November 1941, GMR/T–78/324/6279039–40. The figure of 60,000 tons of supplies given as lost by Becker (*Hitler's Naval War*, London 1974, p. 241) is totally unrealistic.

51 Playfair, *op. cit.*, iii, 107.

52 This is the most recent estimate in Gabriele, *loc. cit.*, p. 287. From his figures it would seem that the small tonnage put across was due less to losses than to the fact that lack of fuel oil forced the Italians to cut the shipping employed by one third.

53 KTB/DAK/Abt. Qu, 25, 26 November 1941, GMR/T–314/16/000012–13.

54 *Ibid*, 27 November 1941, *ibid*, 000014.

55 *Ibid*, 4 December 1941, *ibid*, 000020.

56 *Ibid*, 7, 13, 15 December 1941, *ibid*, 000021–2, 000025–6.

57 *Ibid*, 16 December 1941, *ibid*, 000029.

58 Der dtsch. Gen. b. H.Q. d. Ital. Wehrmacht to OKH/Genst.d.H/Op. Abt, No. 150106/41 g.Kdos, GMR/T–78/324/6279041–42.

59 OKH/Genst.d.H/Op.Abt, No. 36885/41 g.Kdos, Morgenmeldung, 16 December 1941, GMR/T–78/307/6258390.

60 OKW/WFSt/Abt.L to OKH/Genst.d.H/Op.Abt, No. 442070/41 g.Kdos, 5 December 1941, GMR/T–78/324/6279054–55.

61 Funkzentrale Rom to Pz.AOK Afrika/O.Qu, No. 356/41 g.Kdos, 4 December 1941, *ibid*, 6279730.

62 OKH/Genst.d.H/Op.Abt.(IIb) No. 36899/41 g.Kdos, 'Zwischenmeldung von 18 December 1941', GMR/T–78/307/6258539.

63 Der dtsch.. Gen. b. H.Q. d. Ital. Wehrmacht to OKH/Genst.d.H/Op. Abt, Nos. 150113/41 and 150115/41 g.Kdos, 2, 3 December 1941, GMR/T–78/324/6279043–47.

64 Chef OKW/WFSt/Abt. L to dtsch. Gen. b. H.Q. d. Ital. Wehrmacht, No. 442501/41 g.Kdos, 4 December 1941, *ibid*, 6279048–49.

65 U. Cavallero, *Commando Supremo* (Bologna, 1948) p. 160, entry for 8 December 1941.
66 Der dtsch. Gen. b. H.Q. d. Ital. Wehrmacht to OKH/Genst.d.H/Op.Abt, Nos. 150145/41 and 150147/41 g.Kdos, 28, 29 December 1941, GMR/ T–78/324/6279056–62; also Mussolini to Hitler, 28 December 1941, in *Les Lettres secrètes échangées par Hitler et Mussolini* (Paris, 1946) pp. 134–6.
67 Jäckel, *op. cit.*, pp. 207–16.
68 KTB/DAK/Abt.Qu, 30 December 1941, 3 January 1942, GMR/T–314/ 000035–37.
69 The figure for DAK was 25,000 tons; Baum–Weichold, *op. cit.*, p. 212. How much the Italians received can only be roughly calculated on the basis of the shipping tonnage used.
70 KTB/DAK/Abt.Qu, 14, 20, 24 January, 1942, GMR/T–314/16/000042.
71 Der dtsch. Gen. b. H.Q. d. Ital. Wehrmacht to OKH/Genst.d.H/Op.Abt, No. 5001/41 g.Kdos, 7 December 1941, GMR/T–78/324/6279063–6.
72 OKH/Genst.d.H/Gen.Qu.1/I to Op.Abt, 'Ferngespräch Qu. Rom. . .von 18.1.1941', *ibid*, 6279240–41.
73 OKH/Genst.d.H/Gen.Qu. 1 to Qu. Rom, No. I/591/42 g.Kdos, 29 January 1942, *ibid*, 6279242.
74 KTB/DAK/Abt.Qu, 9, 10, 11, 12, 13 February 1942, GMR/T–314/16/ 000055–57.
75 To his wife, however, Rommel complained that he did not have enough trucks; *The Rommel Papers*, p. 186.
76 Der Dtsch. Gen. b. H.Q. d. Ita. Wehrmacht to OKH/Genst.d.H/Op. Abt, No. 150115/41 g.Kdos, 3 December 1941, GMR/T–78/324/ 6279045–47.
77 For detailed figures cf. R. Bernotti, *Storia della guerra nel Mediterraneo* (Rome, 1960) p. 225.
78 Rommel later admitted that 60,000 tons did in fact cover his needs; *The Rommel Papers*, p. 192. However, this figure apparently does not include Luftwaffe supplies.
79 Cavallero, *op. cit.*, pp. 253, 256, entries for 5, 15 May 1942.
80 *Ibid.*, pp. 243–5, entry for 9 April 1942. In his memoirs, Rommel wrote that the supply problem could have been solved 'if the responsible post in Rome had been occupied by a man with enough authority and initiative to tackle it'. *The Rommel Papers*, pp. 191–2.
81 *KTB*/OKW, ii, 1, 324, entry for 18 April 1941.
82 Cavallero, *op. cit.*, pp. 233–4, entry for 17 March 1942.
83 *Ibid*, 234–6, entries for 21, 23 March 1942.
84 *Ibid*, 250–1, entries for 30 April, 9 May 1942; *KTB*/OKW, ii, 1, 331, entries for 1, 7.5.1942.
85 See S. W. Rosskill, *The War at Sea 1939–1945* (London, 1956) ii, 45–6; and A. Kesselring, *Memoirs* (London, 1953) p. 124.
86 Marine Verbindungsoffizier zum OKH/Genst.d.H, No. 31/42 g.Kdos, 10 April 1942, GMR/T–78/646/000985–90; OKH/Genst.d.H/Op.Abt, No. 420169/41 g.Kdos, 9 April 1942, *ibid*, 000964–65.
87 Der dtsch. Gen. b. H.Q. d. Ital. Wehrmacht to OKH/Genst.d.H/Op.Abt, No. 23/42 g.Kdos, 11 June 1942, GMR/T–78/324/627085.

88 This was Rommel's own estimate; *The Rommel Papers*, p. 191. Even so, however, no strategic victory could be expected, but only 'the elimination of the threat from the south for a long time'.

89 *KTB*/Halder, ii, 150, entry for 25 October 1940.

90 *KTB*/OKW, ii, 1, pp. 443–4, ed.'s note (Warlimont).

91 Pz. AOK Afrika to OKH/Genst.d.H/Op.Abt, 22 June 1942, GMR/T–78/324/6279032.

92 Gabriele, *loc. cit.*, p. 287; Playfair, *op. cit.*, iii, 327; Cavallero, *op. cit.*, p. 283, entry for 29 June 1942; der dtsch. Gen. b. H.Q. d. Ital. Wehrmacht to Pz.AOK Afrika, No. 24/42 g.Kdos, 19 June 1942, GMR/T–78/324/6279068.

93 W. Warlimont, 'The Decision in the Mediterranean 1942', in H. A. Jacobsen & J. Rohwehr, *Decisive Battles of World War II* (London, 1965) pp. 192–3. Also Mueller-Hillebrand, *Das Heer*, ii, 86.

94 Cavallero, *op. cit.*, p. 279, entry for 25 June 1942; Kesselring, *op. cit.*, p. 124.

95 Hitler to Mussolini, 21 June 1942, *Les Lettres secrètes*. . .pp. 121–3.

96 Figures from D. Macintyre, *The Battle for the Mediterranean* (London, 1964) p. 146.

97 Pz.AOK Afrika to OKH/Genst.d.H/Op.Abt, No. 3914, 4 July 1942, GMR/T–78/325/6280549.

98 Westphal, *op. cit.*, p. 167.

99 Warlimont, *loc. cit.*, p. 192.

100 Cavallero, *op. cit.*, p. 296, entry for 26 July 1942. Of this amount, 30,000 tons were fuel and 30,000 were meant for DAK; see E. Faldella, *L'Italia e la seconda guerra mondiale* (Bologna, 1959) p. 286, and F. Baylerlein, 'El Alamein', in S. Westphal ed., *The Fatal Decisions* (London, 1956) p. 87.

101 R. Maravigna, *Come abbiamo perduto la guerra in Africa* (Rome, 1949) pp. 354–6.

102 Gabriele, *loc. cit.*, p. 287; Playfair, *op. cit.*, iii, 327; Pz.AOK Afrika to OKH/Genst.d.H/Op.Abt, Nos. 3982, 4103, 4354, 7, 10, 19 July 1942, GMR/T–78/325/6280453, 6280449, 6280434–35.

103 *The Rommel Papers*, p. 234; 'Beurteilung der Lage und des Zustandes der Panzerarmee Afrika am 21.7.1942', in *KTB*/OKW, ii, 1, pp. 515–16.

104 Playfair, *op. cit.*, iii, 327.

105 C. Favagrossa, *Perchè perderemo la guerra* (Milan, 1947) p. 179.

106 Pz..AOK Afrika to OKH/Genst.d.H/Op.Abt, unnumbered, 21 August 1941, GMR/T–78/325/6280384–86.

107 Cavallero, *op. cit.*, 314, 326, entries for 20 August, 7 September 1942; Kesselring, *op. cit.*, pp. 130–1.

108 *The Rommel Papers*, p. 230.

109 Pz.AOK Afrika to OKH/Genst.d.H/Op.Abt, No. 2100 g.Kdos, 21 September 1942, GMR/T–78/325/6280261.

110 This was apparently due to Rommel's successful demand that the Italians cancel their 'idiotic' measure of closing down the port of Tobruk; Pz.AOK Afrika to OKH/Genst.d.H/Op.Abt, Nos 7081, 9138, 6, 16 October 1942, *ibid*, 6280504, 6280519.

111 The figures are: June 13,581 tons, July 11,611, August 45,668, September 15,127, October 32,572; Gabriele, *loc. cit.*, p. 287.

112 The figures are: June 135,847, July 274,337, August 253,005, September 205,559, October 197,201; as against 393,539 in May. *Ibid.*

113 Figures collated from Bragadin, *op. cit.*, pp. 364–7.

114 Cavallero, *op. cit.*, p. 308, entry for 12 October 1942.

115 E.g. Pz.AOK Afrika to OKH/Genst.d.H/Op.Abt, No. 1794/42 g.Kdos, 29 August 1942, GMR/T–78/325/6280370–72.

116 Same to Same, No. 8501, *ibid*, 6280530–31.

117 According to the best information available, only 15 per cent of the supplies, 8.5 per cent of the personnel, and 8.4 per cent of the shipping sent from Italy to Libya in 1940–3 were lost at sea; Bernotti, *op. cit.*, p. 272, and Gabriele, *loc. cit.*, p. 300.

118 Stata Maggiore Esercito/Ufficio Storico ed., *Terza Offensiva Britannica in Africa Settentrionale* (Rome, 1961) p. 300.

119 The figures are: February–June, 22,264 tons per month; July–October, 22,442 tons. Based on Bragadin, *op. cit.*, pp. 154, 287.

120 See footnote 116 *supra*.

121 Pz.AOK/Ia, 86/42 g.Kdos, 'Auszug aus Beurteilung der Lage und des Zustandes der Panzerarmee Afrika am 15.8.1942', GMR/T–78/45/6427948–50.

122 *The Rommel Papers*, p. 328.

Chapter 7

1 This involved the doubling of the speed of all trains; Gröner, *Lebenserinnerungen*, p. 132.

2 C. Barnett, *The Desert Generals* (London, 1961) p. 104ff.

3 *Encyclopaedia Britannica*, 'Logistics', by R. M. Leighton.

4 Napoleon often said he used to plan his campaigns months, even years in advance, but this was never true except possibly in the case of his disastrous war against Russia. Moltke declared that the plan of operations ought not to proceed beyond the first encounter with the enemy. Schlieffen ignored this wise dictum to his cost. Hitler used to say he acted 'with the certainty of a sleepwalker'.

5 It has been estimated that, for every plan put into practice, 20 were discarded; F. Morgan, *Overture to Overlord* (London, 1950) p. 282.

6 W. F. Ross and C. F. Romanus, *The Quartermaster Corps; Operations in the War against Germany* (Washington, D.C., 1965) p. 256.

7 D. D. Eisenhower, *Crusade in Europe* (London, 1948) p. 185.

8 B. H. Liddell Hart, 'Was Normandy a Certainty?', *Defense of the West* (London, 1950) pp. 37–44.

9 G. A. Harrison, *Cross Channel Attack* (Washington, D.C., 1951) p. 54ff.

10 For a good discussion of these factors see Brown, *op. cit.*, pp. 121–26.

11 K. G. Ruppenthal, *Logistical Support of the Armies* (Washington, D.C., 1953) i, 288.

12 Cf. H. Ellis, *Victory in the West* (London, 1953) i, 479–80.

13 Ruppenthal, *op. cit.*, i, 467ff.

14 SHAEF/G–4, 'Weekly Logistical Summary, D−D+11', 24 June 1944, Public Record Office (PRO)/WO/171/146, appendix 3.

15 Ruppenthal, *op. cit.*, i, 469.

16 For the American forces see *ibid*, 447; for the British, 'Second Army Ammunition Holdings, D+6−D+45', the Liddell Hart Papers (States House, Medmenham, Bucks) file 15/15/30.

17 E. Busch, 'Quartermaster Supply of Third Army', *The Quartermaster Review*, November–December 1946, pp. 8–9.

18 The Allied order of battle in north–western France consisted of Montgomery's 21. Army Group on the left and Bradley's 12. Army Group on the right. The former was made up, again from left to right, of Crerar's 2. Canadian and Dempsey's 1. British Army; the latter, of Hodges' 1. and Patton's 3. Armies.

19 J. Bykofsky & H. Larson, *The Transportation Corps* (Washington D.C., 1957) iii, p. 239.

20 By some accounts the difference was as much as fifty per cent; SHAEF/G–4, report of 10 September 1944, *ibid*, WO/219/3233.

21 Annex to SHAEF/G–4/106217/GDP, 'Topography and Communications [in France]', 17 June 1944, PRO/WO/171/146, appendix 1.

22 These figures are given by J. A. Huston, *The Sinews of War* (Washington, D.C., 1966) p. 530. Other sources give different estimates, e.g. the 560 ton 'SHAEF figure' mentioned by Ross & Romanus, *op. cit.*, p. 401. 800 tons per day were taken as a basis for planning 'Overlord'. Of course, all such figures mean little.

23 Figures from Ruppenthal, *op. cit.*, i, 482, footnote 4. GTR companies consisted of forty trucks with an overall capacity of 200 tons and were subordinated to the quartermaster-general for work in the Communications Zone

24 Ross & Romanus, *op. cit.*, p. 475.

25 Thus, throughout the month of August 3. Army lost only a few thousand gallons of fuel to air attack; Busch, *loc. cit.*, p. 10.

26 C. Wilmot, *The Struggle for Europe* (London, 1952) chs. 14 and 27; B. L. Montgomery, *Memoirs* (London, 1958) p. 280ff.

27 Eisenhower, *op. cit.*, pp. 344–5.

28 The logistic argument is most clearly presented by R. G. Ruppenthal in K. R. Greenfeld ed., *Command Decisions* (London, 1960) ch. 15. Other books on the problem are too numerous to be listed here.

29 B. H. Liddell Hart, *History of the Second World War*, p. 584ff.

30 Since 3. Army did not keep any records of deliveries no definite figures are available.

31 Nevertheless, from 23 August to 16 September Hodges received eighty-eight per cent of his requirements in fuel, 'First [US] Army History', Liddell Hart Papers, file 15/15/47.

32 Busch, *loc. cit.*, p. 76.

33 H. Essame, *Patton the Commander* (London, 1974) p. 192.

34 2. [British] Army to SHAEF/G–4, No. 215/30, 30 August 1944, PRO/WO/219/259.

35 *Administrative History 21. Army Group* (London, n.d.) p. 47; also Ellis, *op. cit.*, i, 473.

36 Q.M. Rear H.Q. 21. Army Group telegrams Nos. 18483 and 18489, 1, 16 September 1944, PRO/WO/171/148, appendices 3, 9.

37 This was the SHAEF estimate; Eisenhower, *op. cit.*, p. 312.

38 'State of [German] divisions on Crossing the Seine', Liddell Hart Papers, file 15/15/30.

39 Model to Rundstedt, 27 September 1944, printed in War Office ed., 'German Army Documents Dealing with the War on the Western Front from June to October 1944' (n.p., 1946). Even worse figures are given elsewhere, e.g. Westphal, *Erinnerungen*, p. 277, or Liddell Hart, *History of the Second World War*, p. 585.

40 Eisenhower to Army Group Commanders, 4 September 1944, in A. D. Chandler ed., *The Papers of Dwight David Eisenhower* (Baltimore, Md., 1970) iv, 2115; unsigned, undated memorandum (probably Bedell Smith, 1 September 1944), SHAEF/17100/18/ps (A), 'Advance to the Siegfried Line', PRO/WO/219/260.

41 See Ellis, *op. cit.*, i, 82.

42 The earliest unprinted reference to Montgomery's proposal that I could find is 21. A.G./20748/G(Plans), 'Grouping of the Allied Forces for the advance into Germany', 11 August 1944, PRO/WO/205/247; the earliest printed one, H. C. Butcher, *Three Years with Eisenhower* (London, 1946) p. 550, entry for 14 August 1944.

43 Montgomery, *op. cit.*, p. 266.

44 *Ibid*, 268–9.

45 This was a considerable victory for Montgomery, for the Americans at first wanted to give him one corps only; Bedell Smith memorandum, 22 August 1944, PRO/WO/219/259.

46 Eisenhower to Montgomery, 24 August 1944, in Chandler, *op. cit.*, iv, 2090.

47 Montgomery, *op. cit.*, pp. 271–2.

48 21. AG to SHAEF/G–4, No. D/109/15, undated (9 September 1944?), PRO/WO/205/247. Montgomery's subsequent denial (*op. cit.*, pp. 294–5) that he had used this phrase is false.

49 Eisenhower to Montgomery, 22 September 1944, in Chandler, *op. cit.*, iv, 2175.

50 Montgomery, *op. cit.*, pp. 282–90; Eisenhower, *op. cit.*, pp. 344–5.

51 Cf. J. Ehrman, *Grand Strategy* (London, 1956) v, 379–80.

52 Eisenhower, *op. cit.*, p. 306; O. N. Bradley, A *Soldier's Story* (London, 1951) p. 400.

53 Eisenhower to Montgomery, 13 September 1944, PRO/WO/19/260; SHAEF/G–4, 'Summary of British Rail Position as on 24.9.1944', *ibid*, WO/219/3233; 'British Supplies North of the Seine', Liddell Hart Papers, file 15/15/48.

54 A figure of 130 miles per day was used by the Allied planners who considered an advance on Berlin; SHAEF/G–3, 'Logistical Analysis of Advance into Germany', 6 September 1944, PRO/WO/219/2521.

55 'Supply Problems 21. AG Sept. 1944', signed by Col. O. Poole, 14 September 1944, Liddell Hart Papers, file 15/15/48.

56 Based on a turnaround time over 360 miles of three days.

57 See G. S. Patton, *War as I Knew It* (London, n.d.) p. 128.

58 Based on the fact that, during the Arnhem operation, Eisenhower did furnish Montgomery with 500 tons per day by grounding three American divisions.

59 Lack of detailed information prevents us from taking into accounts such factors as the different kinds of cargo comprising an army's supplies, the state of the roads, traffic control, rates of vehicle attrition, and so on. Our calculations, however, are quite as detailed as the ones on which the planners of SHAEF (e.g. as mentioned in note 54) based themselves; and they are more detailed by far than anything else I have seen printed on the subject.

60 Ruppenthal, *op. cit.*, i, 552, ii, 14. In spite of the initial delay, rail transportation in September came up to the planners' original expectations (2,000,000 ton/miles a day) as laid down in SHAEF/G–4 No. 1062/7/GDP, 'Post "Neptune" Operations – Administrative Appreciation', 17 June 1944, PRO/WO/171/146.

61 Once at the end of July, when the capture of the Brittany ports was postponed, and again in the middle of August, when the decision to cross the Seine was made; cf. Greenfeld, *op. cit.*, pp. 322–3.

62 That is, apart from the American trucks and airplanes already committed in support of the Arnhem operation.

63 As it was, Montgomery was complaining that Hodges did not receive fuel; Bradley to Eisenhower, 12 September 1944, PRO/WO/219/260. In fairness to the Americans it must be added that Montgomery refused to do a thing to help them; 12. AG report, 4 September 1944, *ibid*, WO/219/2976.

64 For list of ports used by the British cf. 'Musketeer', 'The Campaign in N.W. Europe, June 1944–February 1945', *Journal of the Royal United Services Institute*, 1958, p. 74.

65 Montgomery, *op. cit.*, p. 272.

66 'Supply Problems 21. AG, Sept. 1944', signed by Col. O. Poole, Liddell Hart Papers, file 15/15/48. According to another scheme, three airborne divisions were also to be used; 'Allied Strategy after Fall of Paris' (interview by Chester Wilmot of General Graham, formerly MGA 21. AG, London, 19 January 1949), *ibid*.

Chapter 8

1 A. C. P. Wavell, *Speaking Generally* (London, 1946) pp. 78–9.

2 *Encyclopaedia Britannica*, 14th ed., 1973, 'Logistics', by R. M. Leighton.

3 Behr, *op. cit.*, pp. 18–23.

4 Huston, *op. cit.*, p. 673, concludes that the speed of strategic marches had not risen significantly, as does E. Muraise, *Introduction à l'Histoire militaire* (Paris, 1964) p. 210. On the other hand, Basil Liddell Hart believed that 'it is absolutely untrue that the advances in World War II by mechanized forces were remarkably similar' to those of earlier days; Liddell Hart to H. Pyman, 28 November 1960, Liddell Hart Papers, file 11/1960/7.

5 See the interesting remarks in J. Wheldon, *Machine Age Armies* (London, 1968) pp. 172–3, 179.

Postscript

1 See, above all, B. von Baumann, *Studien ueber die Verpflegung der Kriesheere in Felde*, 3 vols., Leipzig, 1866–80.
2 Field Marshal B. M. Montgomery of Alamein, *From Normandy to the Baltic*, Boston, Mass., 1948, p. 145.
3 Good bibliographies on the subject may be found in J. A. Lynn, ed., *Feeding Mars: Logistics in Western Warfare from the Middle Ages to the Present*, Boulder, Colo., 1993, pp. 289–308; also H. Roos, ed., *Van marketenstrer tot logistiek netwerk; de militaire logistiek door de eewen heen*, Amsterdam, 2002, pp. 391–404.
4 On the way the term was expanded, see K. Macksey, *For Want of a Nail; The Impact on War of Logistics and Communications*, London, Brassey's, 198, p. 5.
5 B. S. Bachrach, 'Logistics in Pre-Crusade Europe', in Lynn, ed., *Feeding Mars*, pp. 60–8.
6 According to K. Hopkins, 'The Transport of Staples in the Roman Empire', in P. Garnsey and C. R. Whittaker, *Trade and Staples in Antiquity (Greece and Rome)*, Budapest, Akademiai Kiado, 1987, p. 86 table 2.
7 J. Roth, *The Logistics of the Roman Army at War*, Leiden, 1999, pp. 163, 174, 319.
8 *Ibid*, p. 174.
9 C. von der Goltz, *The Nation in Arms*, London, 1913, p. 457.
10 For a critique of my calculations in this respect see J. A. Lynn, 'The History of Logistics and Supplying War', in Lynn, ed., *Feeding Mars*, especially p. 22 and table 2.1. Here, I do not want to deal with Lynn's critique in detail. However, one wonders what to make of an author who, adding up 7.17, 3.64, 11.40, 0.43 and 7.28 (all these are pounds of supplies per man per day) gets 66.8 pounds, instead of 29.92 which is the correct answer.
11 J. A. Huston, *The Sinews of War: Army Logistics 1775–1953*, Washington, D.C., 1966, p. 398.
12 See on this, most recently, L. Sukstorf, *Die Problematik der Logitik im duetshcen Heer waehrend des deutsch-franzoesischen Krieges 1870/71*, Frankfurt/Main, 1994, pp. 143–45.
13 See on this W. H. Pagonis, *Moving Mountains: Lessons in Leadership and Logistics from the Gulf War*, Cambridge, Mass., 1991, p. 106.
14 J. Shy, 'Logistical Crisis and the American Revolution: A Hypothesis', in Lynn, ed., *Feeding Mars*, pp. 171, 173.
15 Livy, *Roman History*, London, Loeb Classical Library, 1938, 22.11.5.
16 F. E. Adcock, *The Greek and Macedonian Art of War*, Berkeley, Calif., 1957, p. 65.
17 D. W. Engels, *Alexander the Great and the Logistics of the Macedonian Army*, Berkeley, Calif., 1978, particularly pp. 121–2.
18 E.g. Herodian, London, Loeb Classical Library, 1969, 8.2.6–5.6; Polybios, *The Histories*, London, Loeb Classical Library, 1922–, 1.18.10; Livy, *Roman History*, 34.34.2–6 (quoting Titus Quinctius Flamininus).
19 M. van Creveld, *Command in War*, Cambridge, Mass., 1985, pp. 58–102.
20 Huston, *The Sinews of War*, p. 222.

21 According to E. Hagerman, *The American Civil War and the Origins of Modern Warfare*, Bloomington, Ind., Indiana University Press, 1988, p. xv.

22 Quoted in Macksey, *For Want of a Nail*, p. 21.

23 *Command in War*, pp. 158–60.

24 See, e.g., P. M. S. Bell, *The Military and Political Consequences of Atomic Energy*, London, 1948, chapter 10; also J. F. C. Fuller, *The Conduct of War since 1789*, London, 1961, pp. 321 ff.

25 For a short account of the decline of conventional war as perceived by this author, see M. van Creveld, *The Rise and Decline of the State*, London, 1999, pp. 337–54.

26 See on this entire subject Th. Kane, *Military Logistics and Strategic Performance*, London, 2001; and N. Brown, *Strategic Mobility*, London, 1963, which, in spite of its age, remains by far the best of its kind.

27 Pagonis, *Moving Mountains*, p. 7.

28 Pagonis, *Moving Mountains*, p. 146.

29 D. M. Moore, J. P. Bradford and P. D. Antill, 'Learning from Past Experience: Is What Is Past a Prologue?' *Whitehall Paper*, London, 2000, pp. 51–2.

30 See on this Pagonis, *Moving Mountains*, pp. 151–58.

31 Moore, Bradford and Antill, 'Learning from Past Experience', p. 59.

32 Pagonis, *Moving Mountains*, p. 204.

33 See on this, in general, C. S. Gray, *Strategy for Chaos: RMA and the Evidence of History*, London, 2002.

34 See on this Larry Haukens, 'Agile Logistics' (2001), available at <http://log.dau.mil/papers/research/apwc%20200/larry%20haukens/doc>

35 See *Supplying War*, p. 167.

36 J. S. Miseli, 'The View from My Windshield: Just-in-Time Logistics Just Isn't Working', *Armor*, 112, 5, September–October 2003, pp. 11–12.

37 Some discussion of this subject may be found at <www.tank-net.org>; see also A. Cordesman, 'The Instant Lessons of the Iraq War', 28 March 2003, p. 130, available at <www.csis.org>. As usual, I have Zeev Elron to thank for these references.

38 According to Huston, *The Sinews of War*, p. 398.

39 There is a short summary of the various generations in W. S. Lind, 'The Four Generations of War', *Counterpunch*, 23 April 2003, available at <http://www.counterpunch.org/lind04232003.html>.

40 C. von Clausewitz, *On War*, M. Howard and P. Paret, eds., Princeton, N.J., 1976, pp. 119–22.

INDEX